JUNE 2021

CLIMATE CHANGE AND HUNGER

ESTIMATING COSTS OF ADAPTATION IN THE AGRIFOOD SYSTEM

TIMOTHY B. SULSER, KEITH WIEBE, SHAHNILA DUNSTON, NICOLA CENACCHI, ALEJANDRO NIN-PRATT, DANIEL MASON-D'CROZ, RICHARD ROBERTSON, DIRK WILLENBOCKEL, AND MARK W. ROSEGRANT

A Peer-Reviewed Publication
International Food Policy Research Institute
Washington, DC

ABOUT IFPRI

The International Food Policy Research Institute (IFPRI), a CGIAR center established in 1975, provides research-based policy solutions to sustainably reduce poverty and end hunger and malnutrition. IFPRI's strategic research aims to foster a climate-resilient and sustainable food supply; promote healthy diets and nutrition for all; build inclusive and efficient markets, trade systems, and food industries; transform agricultural and rural economies; and strengthen institutions and governance. Gender is integrated in all the Institute's work. Partnerships, communications, capacity strengthening, and data and knowledge management are essential components to translate IFPRI's research from action to impact. The Institute's regional and country programs play a critical role in responding to demand for food policy research and in delivering holistic support for country-led development. IFPRI collaborates with partners around the world.

Photo credits: (cover) Petterik Wiggers / Panos Pictures

ISBN: 978-0-89629-416-5

DOI: https://doi.org/10.2499/9780896294165

CONTENTS

TABLES

FIGURES

APPENDIXES

ACKNOWLEDGMENTS

We would like to thank the co-authors of the Rosegrant et al. (2017) study and other developers of the IMPACT model (Robinson et al. 2015), who all contributed to the foundations of this analysis through collaborative modeling and foresight analysis work. We also thank the participants of the expert technical consultation (Improving Cost Data and Methods for Evaluating the Cost-Effectiveness of Agriculture and Food Systems Interventions, held June 18, 2019) who made important contributions to our methodological review. Funding for the current analysis was provided by the Bill & Melinda Gates Foundation (BMGF), but this work also builds on previous work supported by the CGIAR Research Program on Policies, Institutions, and Markets led by IFPRI; the CGIAR Research Program on Climate Change, Agriculture and Food Security; the US Agency for International Development; and previous support from BMGF for the Global Futures and Strategic Foresight program.

ABSTRACT

This report assesses the cost of adaptation to climate change across a range of future climate scenarios and investment options. We focus on offsetting climate change impacts on hunger through investment in agricultural research, water management, and rural infrastructure in developing countries. We link climate, crop, water, and economic models to (1) analyze scenarios of future change in the agriculture sector to 2050 and (2) assess trade-offs for these investments across key Sustainable Development Goals (SDGs) for poverty, hunger, and water. Our reference projections show that climate change slows progress toward eliminating hunger, with an additional 78 million people facing chronic hunger in 2050 relative to a no-climate-change future, over half of them in Africa south of the Sahara. Increased investments can offset these impacts. Achieving this would require that annual investment in international agricultural research increase from US$1.62 billion to US$2.77 billion per year between 2015 and 2050. Additional water and infrastructure investments are estimated to be more expensive than agricultural R&D at about US$12.7 billion and US$10.8 billion per year, respectively, but these address key gaps to support transformation toward food system resiliency. Findings on ranges of costs and trade-offs and complementarities across SDGs will help policymakers make better-informed choices between alternative investment strategies.

INTRODUCTION

Climate change is a growing challenge for food systems around the world, presenting both changing trends and increased variability in temperatures and precipitation. These changes threaten agricultural productivity, food and nutrition security, health, and the potential for achieving key Sustainable Development Goals (SDGs) for poverty, hunger, and water, among others. Despite these challenges, however, appropriate investments in the agriculture sector can help people and food systems adapt by increasing productivity, resilience, and resource-use efficiency. To help policymakers and other decision-makers evaluate alternative adaptation strategies, this report assesses the cost of adaptation to climate change across a range of future climate scenarios and investment options, with a particular focus on adaptation to offset climate change impacts on hunger.

Climate change, its impacts on agriculture, and measures to address those impacts are all highly complex phenomena. Estimating the cost of such measures is correspondingly difficult. Rather than attempting to estimate the total cost of adaptation, Nelson et al. (2009) focused on one particular dimension of these phenomena: the cost of selected agricultural investment strategies to offset the adverse impact of climate change on humans as proxied by future projections of childhood undernutrition. The assumption was that if we can avert impacts on the most vulnerable, then we will likely have taken care of the rest of the population as well. This report does the same, but with an updated methodology and a broader focus on an estimate of the total population at risk of chronic hunger. We draw on recent analyses by Rosegrant et al. (2017), Rosegrant et al. (2021), and Mason-D'Croz et al. (2019), which in turn built on Nelson et al. (2009).

Other studies have used a variety of approaches to estimate the cost of climate change adaptation. Country- and sector-level studies in the 1990s assessed adaptation costs based on the estimated damages from climate change impacts (Tol 2005; Tol, Fankhauser, and Smith 1998). The first global estimates of adaptation costs shared a methodology developed by the World Bank (2006) that estimated the fraction of investments flowing to developing countries deemed sensitive to climate change impacts and then used a "markup" factor to represent the costs of "climate-proofing" those investments.[1] These studies did not specify a particular time horizon or climate

1 The initial analysis estimated that 2–10 percent of gross domestic investment, 10 percent of foreign direct investment, and 40 percent of official development assistance were sensitive to climate change. The authors estimated that a markup of 10–20 percent would be needed to climate-proof these investments (World Bank 2006).

scenario, but the use of current investment flows as a starting point suggests that they focused on short-term adaptation costs.

Estimates vary widely due to the lack of empirical data on climate sensitivity and the costs of climate-proofing. Studies by UNFCCC (2007) and Forstater and Zadek (2009) specified a time horizon of 2030 and included more detailed assessments of investment and financial flows for each sector. A 2010 review of published studies on the global cost of adaptation found that estimates ranged between US$25 and US$100 billion per year for the period 2015–2030 (Agrawala and Fankhauser 2008; Fankhauser 2010). These figures referred to the costs of adapting to a warming of 2–3 degrees Celsius and were estimated to rise exponentially for higher temperatures. A subsequent analysis by the World Bank (2010) estimated the cost at US$70–100 billion annually for developing countries for the period 2010 to 2050. These were the most recent comprehensive estimates as of the Intergovernmental Panel on Climate Change's (IPCC) Fifth Assessment Report (Chambwera et al. 2014), and arguably the most robust (O'Garra and Mourato 2016). A recent study by Baldos, Fuglie, and Hertel (2020) focused on national public spending on R&D at aggregate regional levels; they estimated that climate adaptation costs that offset climate impacts on crop yields would cost a total of US$187 to US$1,384 billion from 2020 to 2040.

In contrast to these studies, Nelson et al. (2009) focused more specifically on a detailed analysis of the impacts of climate change on: (1) agricultural productivity and related outcomes (including child undernourishment) to 2050, by country and commodity; and (2) the cost of selected agricultural investment strategies in developing countries (namely agricultural research, soil and water management, and rural infrastructure) to counter those impacts. More recent work by Rosegrant et al. (2017) and Mason-D'Croz et al. (2019) extended that approach, incorporating updated models and data and an emphasis on looking at trade-offs and complementarities across different investment strategies. Additional analyses led by colleagues from across the CGIAR broadened these two studies by taking deeper looks at particular commodities and development issues, such as fish, livestock, cereals, and roots, tubers, and bananas (Chan et al. 2019; Enahoro et al. 2019; Kruseman et al. 2020; Petsakos et al. 2019), and agriculture sector income and employment (Frija et al. 2020).

APPROACH AND METHODS

MODELING FOOD SYSTEM FUTURES UNDER CLIMATE CHANGE AND ADAPTATION INVESTMENT SCENARIOS

This analysis builds on the recent work by Rosegrant et al. (2017) and others by analyzing a more complete range of future scenarios along with sensitivity simulations on key costing parameters. The goal is to help establish a more robust foundation for the assessment of the cost of adaptation to climate change for the food system in the developing world. The core output of this study is a direct analog and update to the cost of climate change adaptation presented in Nelson et al. (2009). See Appendix A for the core details for comparison.

To simulate changes in the global food system and assess investment scenarios for their effectiveness to offset climate change impacts out to mid-century, we used IFPRI's International Model for Policy Analysis of Agricultural Commodities and Trade (IMPACT) (Robinson et al. 2015). This modeling framework links climate, crop, water, and economic models to analyze scenarios of future change in agricultural production, consumption, prices, and trade at national, regional, and global scales. The methodology behind the IMPACT modeling framework has evolved and been refined over the past couple of decades in collaboration with other leading research groups around the world, particularly with partners across CGIAR and in the Global Economics group of the Agricultural Model Intercomparison and Improvement Project (AgMIP, https://agmip.org/global-economics/). The IMPACT modeling framework has been applied extensively in numerous peer-reviewed publications and decision-making fora, such as international development organizations, national governments, foundations and grant-making institutions, and across CGIAR itself (https://www.ifpri.org/project/ifpri-impact-model). IMPACT produces reference scenarios to 2050 assuming no climate change as well as different levels of climate change and socioeconomic assumptions from the IPCC, capturing a wide range of possible climate and socioeconomic futures (Table 1). In conjunction with these reference scenarios, we then analyzed scenarios of how plausible investment options – including investments in agricultural R&D, water management, and infrastructure (determined in collaboration with experts across the CGIAR) – could help offset the potential impacts of climate change.

Many complex connections exist between climate change and hunger (Mbow et al. 2019; Beach et al. 2019; Nelson et al. 2018). This analysis focuses on a subset of those. In the most basic sense, climate change, by and large, negatively affects agricultural production by reducing yields. This reduced productivity then reduces availability and drives prices upward, ultimately resulting in increased hunger. The investments modeled here generate economic benefits through several pathways. Specific details are summarized in Appendix D, but in general, increased agricultural R&D boosts crop and livestock yields, reduces food prices, and increases farm income and economywide gross domestic product (GDP) through multiplier effects on the nonagriculture sectors. Increasing incomes, in turn, shift consumers toward different diets, with consumption of usually more food and more diverse food types. Irrigation and water use efficiency investments increase crop yields and reduce prices, and generating higher incomes. Enhanced rural infrastructure reduces postharvest losses and marketing margins, improving the profitability of farm production, and boosting supply to consumers for any given level of production. These effects also increase farm and economywide income, which increases food consumption and reduces hunger. While the investment alternatives modeled in this study are extensive, this list is not exhaustive and the focus in this study is on classes of investments that are commonly recognized across the agricultural development community and that can be modeled directly in IMPACT.

To expand the representation across the range of core drivers, we extended the original scenario set in Rosegrant et al. (2017) to show outcomes across different combinations of climate change assumptions (using Representative Concentration Pathways (RCPs), Global Circulation Models (GCMs), and a no climate change (NoCC) scenario), and scenarios of socioeconomic development (using Shared Socioeconomic Pathways (SSPs)) and agriculture sector investments, for a total of 36 scenarios. Detailed descriptions of RCPs, GCMs, and SSPs are available from Moss et al. (2010), O'Neill et al. (2017), and Navarro-Racines et al. (2020), respectively. These core scenarios are representative of possible futures for the global food system based on the range of core driver assumptions available in the IMPACT model.

This analysis uses the number of people facing chronic hunger as the core indicator of climate change impacts and adaptation. Chronic hunger is calculated based on per capita calorie availability (including access via international trade) and minimum dietary energy requirements, following a methodology equivalent to the Food and Agriculture Organization's (FAO) prevalence of undernourishment (PoU) indicator (details in Appendix D and Robinson et al. 2015). The target for adaptation in this analysis is to offset the effects of climate change by making investments that reduce the number of hungry people projected in 2050 to the same level that would be achieved in the absence of climate change.

We detail here methods for the key advancements of this research; additional core methods used in support of this study are documented in detail in Rosegrant et al. (2017) and Mason-D'Croz et al. (2019).

INVESTMENT COSTS

Costs of investments in agricultural research, water management, and infrastructure are estimated as detailed in Rosegrant et al. (2017) and Mason-D'Croz et al. (2019). The lagged effects of investments in agricultural research are estimated using a perpetual inventory method (PIM) analogous to that used for physical capital (Esposti and Pierani 2003), with supporting data from a variety of recent resources. The underlying assumption behind R&D investment and

TABLE 1 SUMMARY DESCRIPTION OF REFERENCE AND INVESTMENT SCENARIOS

Scenario	Description	Abbreviation
Reference	With no climate change (NoCC; constant 2005 climate with various Shared Socioeconomic Pathways (SSPs))	REF-NoCC*
	With climate change (CC; combinations of SSPs and Representative Concentration Pathways (RCPs) across a range of General Circulation Models (GCMs))	REF-CC*
Agricultural R&D	Increased research and development (R&D) investment across the CGIAR portfolio plus faster and more efficient adoption of new technologies	HIGH+RE
Water Management	Expansion of irrigated area coupled with increased water use efficiency	IX+WUE
	Improved soil water-holding capacity	SWHC
Infrastructure	Infrastructure improvements to improve market efficiency through the reduction of transportation costs and marketing margins (rail, road, port, and electrification)	RMM
Comprehensive	A combination of the Agricultural R&D, Water Management, and Infrastructure scenarios described above	COMP

Note: *In this study, the combination that defines the REF-NoCC scenario is the Shared Socioeconomic Pathway "2" (SSP2) alone, while the REF-CC scenario is the combination of SSP2 with RCP8.5 via the UK Met Office Hadley Centre Earth System Model (HGEM) GCM.

productivity is that a flow of R&D investments creates a stock of knowledge that yields returns into the future. Information needed to construct such a stock includes the flow of gross R&D investment to determine how fast it builds the stock of knowledge, and how fast knowledge becomes obsolete, which is captured by a decay or depreciation rate. Unit cost estimates from existing literature (FAO 2016; Inocencio et al. 2007; IWMI 2012; Rosegrant et al. 2018) are used to calculate investments in water management (irrigation expansion, water use efficiency improvements, and improvements in rainwater harvesting and soil water holding capacity) and infrastructure (improvements to transportation infrastructure and increased rural electrification). With available current data, final cost estimates are evaluated at regional aggregations of East Asia and Pacific (EAP), South Asia (SAS), Africa south of the Sahara (SSA), Middle East and North Africa (MENA), and Latin America and the Caribbean (LAC). Only developing countries are included.

To support this research and review the latest advances in the field, we held a technical consultation (Improving Cost Data and Methods for Evaluating the Cost-Effectiveness of Agriculture and Food Systems Interventions) in June 2019 with a range of subject matter experts in the field of costing investments in the agriculture sector. This helped us focus the sensitivity test of key parameters in the PIM that embody the greatest amount of uncertainty (reference and investment scenario elasticities for international and national investment levels to knowledge stock accumulation and knowledge stock depreciation rates).

TABLE 2 PIM COSTING METHODOLOGY SENSITIVITY SIMULATION SETUP ACROSS CORE R&D COST-RELATED PARAMETERS
(as percentages of calibrated values for the SSP2 reference case)

Cost Sensitivity Simulation	Sensitivity Adjustments		
	Knowledge stock	Elasticities of investment level to knowledge stock accumulation in both national and international agricultural research systems	
	Depreciation rate	Reference scenario	Investment scenarios
All-1	80%	80%	80%
All-2 (BAU)	100%	100%	100%
All-3	120%	120%	120%
DepRate-4	80%		
DepRate-5	120%		
ElasBase-6		80%	
ElasBase-7		120%	
ElasScnr-8			80%
ElasScnr-9			120%

SENSITIVITY TESTING

The primary suggestion from the expert technical workshop was to run sensitivity tests across different key dimensions of our methodology. Part of this was addressed using the 36 different scenarios across the range of IMPACT model drivers that show outcomes for the population at risk of hunger across a range of possible futures for the global food system, as noted above. Next, we conducted sensitivity simulations on key parameter assumptions in the costing models for agricultural R&D (using the PIM) and for water management and infrastructure development (unit costing).

The sensitivity testing of the PIM costing methodology entailed running simulations with combinations of adjustments in three sets of core parameters (Table 2) that together determine the effectiveness and longevity of investments in agricultural R&D. Adjustments of +/−20 percent were made to:

1. elasticities relating investment levels to knowledge stock accumulation in both national and international agricultural research systems in the business-as-usual (BAU) scenario;
2. elasticities relating investment levels to knowledge stock accumulation in both national and international agricultural research systems in the investment scenarios; and
3. rates of knowledge stock depreciation in the BAU and investment scenarios.

Ranges of elasticities across the sensitivity simulations are similar to the bounding exercise conducted by Baldos, Fuglie, and Hertel (2020).

In addition, sensitivity bounds for water and infrastructure investments were incorporated into this analysis. This was accomplished by making adjustments of +/−15 percent in the unit cost assumptions derived from the most up-to-date sources for investments in water management (irrigation expansion, water use efficiency, soil water-holding capacity) and infrastructure (road, rail, port, and electrification improvement/expansion).

The range of reference and investment scenarios and cost sensitivity simulations included in this study enables presentation of a range of results. There is a reference value that corresponds to the calibrated "best" estimate based on available data and expert opinion, and a "plausible interval" that can be interpreted similarly to a confidence interval, so we are highly confident that the true value lies within the interval shown.

For clarity, Box 1 includes an explanation of how we refer to the various scenarios and costing simulations in this study.

BOX 1 REFERRING TO SCENARIOS AND SIMULATIONS IN THIS STUDY

1. The "reference" scenario is the core scenario against which most comparisons are made unless explicitly stated otherwise. In this report, this corresponds to the scenario using driver assumptions drawn from SSP2 and RCP8.5 with the UK Met Office Hadley Centre Earth System Model (HGEM) GCM. In certain cases, we add the term "climate change" for clarity. The abbreviation for this scenario is REF-CC.

2. There is also a "no climate change" version of the reference scenario that uses the SSP2 socioeconomic assumptions and a constant 2005 climate, but this is referred to explicitly in all cases and used only to estimate the impact of climate change. The abbreviation for this scenario is REF-NoCC.

3. In a few situations we need to specify scenarios that either include or do not include climate change but are not associated with the reference; these are referred to as CC and NoCC, respectively.

4. The investment scenarios build upon the reference scenario (with adjustments for alignment with specific SSPs and RCPs) and are referred to explicitly as shown in Table 1 with their appropriate SSP and RCP specifications.

5. The 36 scenarios that were run across the IMPACT model driver dimensions and investment scenarios (listed in Table 1) show ranges of results directly from the IMPACT model; in this case the primary result of interest is the population at risk of hunger, but other results are included to broaden the discussion.

6. The 8 sensitivity simulations run on the PIM costing method (noted in Table 2 along with the reference case "All-2") show the ranges of costing results for the REF-CC and HIGH+RE scenarios. The HIGH+RE costing results are also incorporated in the COMP scenario.

ASSESSMENT OF TRADE-OFFS

Policymakers faced with making investment decisions inevitably must consider trade-offs across multiple dimensions. This work incorporates a range of methods and indicators to enable quantitative assessments of potential trade-offs between competing and complementary multidimensional targets, such as the United Nations SDGs (UN 2015). We focus here on the SDGs for poverty (SDG 1), hunger (SDG 2), and water (SDGs 6 and 12), while further indicators are presented in Rosegrant et al. (2017). The SDGs and specific targets evaluated under the alternative investment trade-offs are:

- SDG 1 – No Poverty, End poverty in all its forms everywhere
  - Target 1.1 – Eradicate extreme poverty

- Target 1.2 – Reduce poverty by at least 50 percent

- SDG 2 – Zero Hunger, End hunger, achieve food security and improved nutrition, and promote sustainable agriculture
  - Target 2.1 – Universal access to safe and nutritious food
  - Target 2.3 – Double the productivity and incomes of small-scale food producers
  - Target 2.4 – Sustainable food production and resilient agricultural practices
- SDG 6 – Clean Water and Sanitation: Ensure access to water and sanitation for all
  - Target 6.4 – Increase water use efficiency and ensure freshwater supplies
  - Target 6.5 – Implement integrated water resources management
- SDG 12 – Responsible Consumption and Production: Ensure sustainable consumption and production patterns
  - Target 12.2 – Sustainable management and use of natural resources

SDG 2 has several targets that overlap with other SDGs, such as Target 2.3, which addresses incomes that are also an integral component of SDG 1, and Target 2.4 on sustainable food production, also addressed under SDG 12. We take a narrow view of SDG 2 to focus on hunger and agricultural production/supply and leave these overlapping targets to their respective SDGs (1, 6, and 12). In the Findings section below, we refer to key components that are explicitly modeled in the IMPACT framework and are aligned with SDG targets. These include: changes in aggregate country income ("GDP") under SDG 1; total agricultural supply ("Ag Supply"), population at risk of hunger ("Hunger") using an equivalent of FAO's PoU calculation, and per capita calorie availability ("Per Capita KCAL Availability") under SDG 2; and blue water use and efficiency measures and irrigation water supply reliability ("Blue Water Use," "Irrigation Water Reliability," and "Blue Water per KCAL") under SDG 12.

FINDINGS

POPULATION AT RISK OF HUNGER ACROSS REFERENCE AND INVESTMENT SCENARIOS IN IMPACT MODEL

In the absence of climate change, the number of people who face chronic hunger would be expected to decline by more than one-half between 2010 and 2050, led by EAP and SAS (Figure 1). Climate change is projected to slow this progress, with an additional 78 million people facing chronic hunger in 2050 (relative to the REF-NoCC scenario) – over one-half of them in SSA. Impacts vary across the range of climate and socioeconomic scenario specifications run in this analysis with high population, low income growth, and rapid climate change being particularly impactful, resulting in an increase above 2010 levels under climate change in all regions except EAP.

Levels of hunger can rise even without climate change in some regions and scenarios, for example, under high population growth coupled with low income growth (SSP3), especially in SSA. In LAC and MENA, hunger rates are low enough that population growth can drive an increase in the number of people at risk of hunger regardless of climate change. Further progress against chronic hunger in LAC and MENA is dependent on drivers outside of IMPACT's food system model of the population at risk of hunger (such as health and education systems).

As modeled here, increased investments in the agriculture sector can more than compensate for climate change effects on the number of hungry people in the world. Improvements in the global situation for the population at risk of hunger under a NoCC future lead to at least an 18 percent reduction in the number of hungry people in 2050 compared to 2010 and, at best, a two-thirds reduction. Under the CC scenarios, the change in the total population at risk of hunger compared to 2010 ranges from an almost 10 percent increase – in the worst situation – to a three-fifths reduction under SSP1.

Comparing the CC and NoCC scenarios in 2050 (instead of in 2010), the increase in the population at risk of hunger in 2050 due specifically to climate change is at least 16 percent, and up to one-third greater. SSA and SAS are the regions where the effects of climate change on the population at risk of hunger vary most widely across socioeconomic and climate scenarios (that is, they have the greatest minimum/maximum ranges among the aggregate regions presented). The range around the reference scenario in SAS is −28 percent to +174 percent, while in SSA it is −48 percent to +94 percent. The COMP investment scenario more than offsets the climate effects modeled here, with an 18 percent to 13 percent reduction in the global population at risk of hunger compared to a future without climate change (NoCC).

It is useful to separate the effects of the various socioeconomic, climate, and investment assumptions underlying the different scenarios. Socioeconomic assumptions play a critical role in scenario outcomes (Figure 2a). Population growth (slowest in SSP1 and fastest in SSP3) is an important driver, but per capita income growth (fastest in SSP1 and slowest in SSP3) is an even stronger force for improving well-being across the different regions. SSP2, which is the "BAU" future used in this study, shows large reductions in hunger across the globe, especially in EAP and SAS, but persistence of hunger in SSA. Under SSP2 the global population at risk of hunger is cut in half between 2010 and 2050. SSP1 is a more favorable set of assumptions, leading to a nearly two-thirds reduction in the global population at risk of hunger over this same period, primarily by reducing incidence in SSA and across Asia. Under SSP3 hunger increases across the globe in comparison with SSPs 1 and 2, tied to slower growth in per capita incomes and accelerated population growth, especially in SSA, which sees a 36 percent increase in the number of hungry people from 2010 to 2050. Compared to

FIGURE 1 POPULATION AT RISK OF HUNGER IN 2010 AND 2050,
globally and by region in the reference (with and without climate change) and investment scenarios

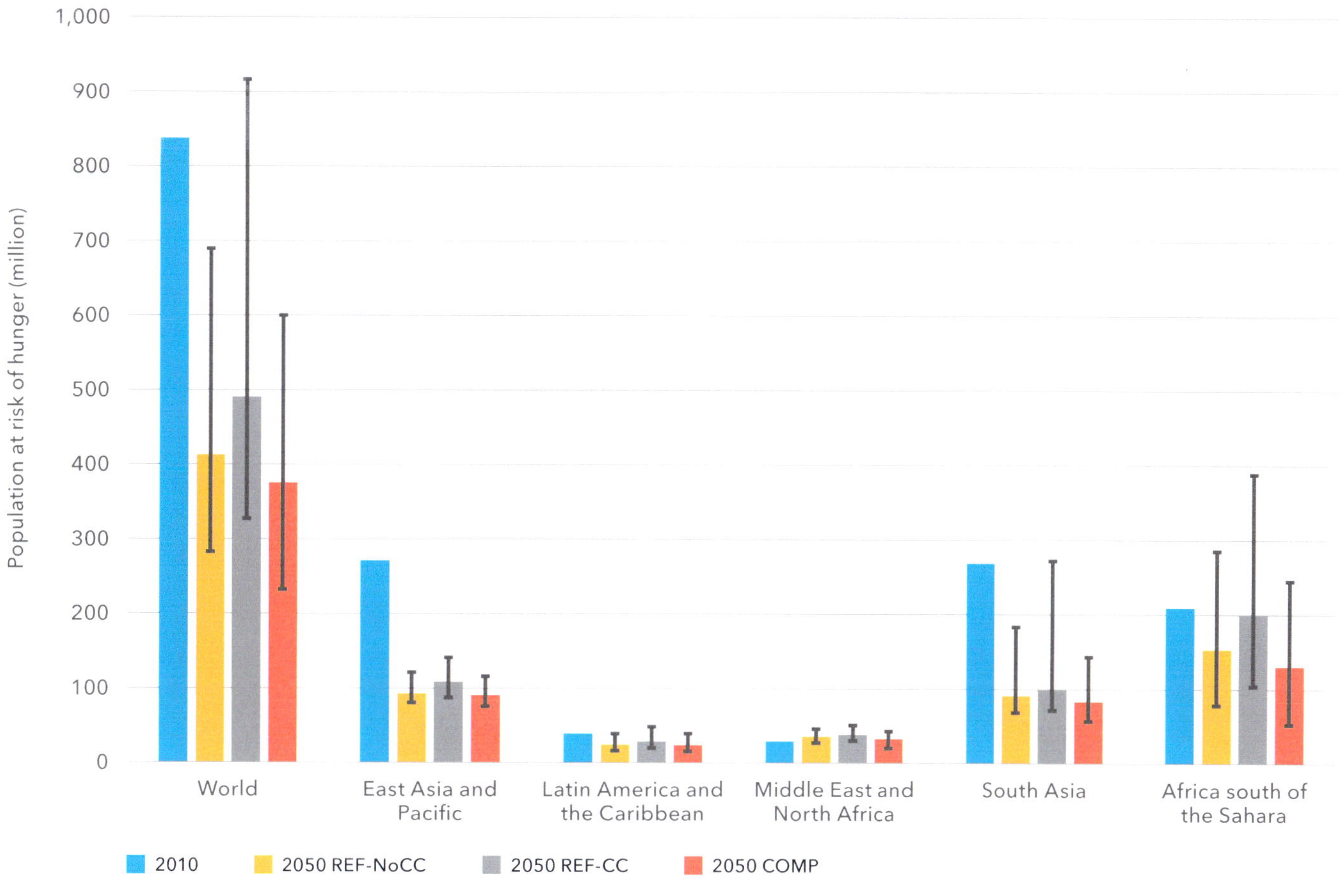

Note: World includes all countries; regions include only developing countries. Data shown are a reference estimate (columns) with the minimum and maximum of a plausible interval (whiskers) using methods described.

SSP2, the socioeconomic conditions in SSP3 lead to 277 million more people at risk of hunger in 2050 while the SSP1 scenario leads to a 130-million-person reduction.

Across the five GCMs used to implement the RCP8.5 emissions pathway (Figure 2b), the global change in population at risk of hunger in 2050 compared to a REF-NoCC scenario in 2050 ranges from a 19 percent increase to a 4 percent decrease (78 million more hungry to 15 million fewer) in UK Met Office Hadley Centre Earth System Model (HGEM) and Geophysical Fluid Dynamics Laboratory (GFDL), respectively. SSA is the region most sensitive to the different conditions encountered across the GCMs, accounting for the majority of change when compared to the REF-NoCC scenario.

Not surprisingly, results also vary significantly across climate pathways. Impacts under RCP4.5 are about half as severe as those for RCP8.5 under SSP2 as modeled with the HGEM GCM (Figure 2c). RCP4.5 sees only a 9 percent increase in the population at risk of hunger, compared to 19 percent under RCP8.5. The impact of RCP8.5 on the number of hungry people ranges from a 16 percent increase to a full one-third increase under SSP1 and SSP3, respectively. Under the socioeconomic conditions in SSP3, the world appears more susceptible to climate change impacts on food security. SAS, for example,

FIGURE 2 POPULATION AT RISK OF HUNGER FROM 2010 TO 2050 ACROSS RANGES OF CORE DRIVER ASSUMPTIONS IN THE IMPACT MODEL

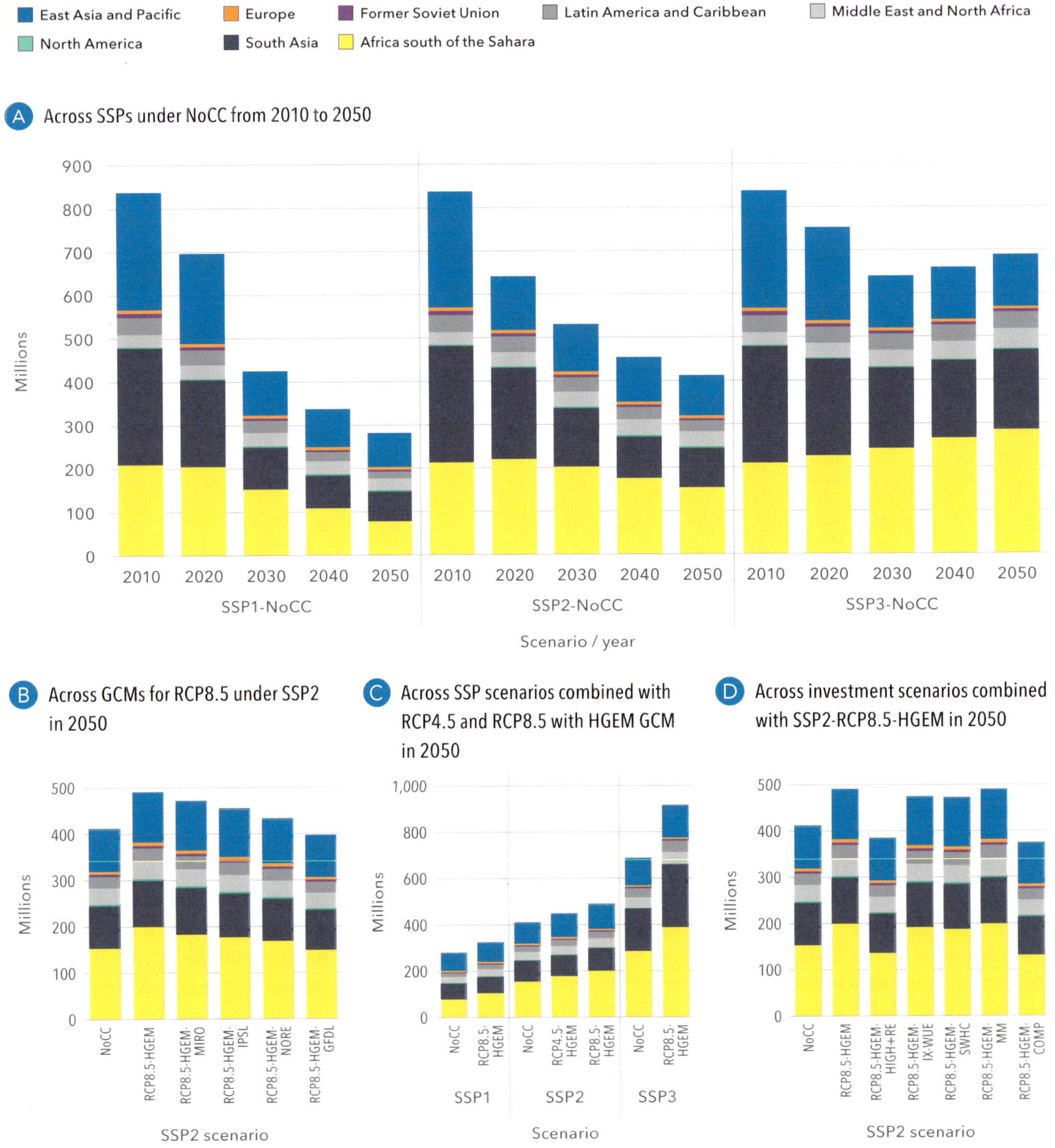

Note: Scales vary by panel. HGEM- UK Met Office Hadley Centre Earth System Model

sees a 48 percent increase in the population at risk of hunger compared to a less than 10 percent increase under SSPs 1 and 2.

Finally, the investment scenarios contribute to offsetting the effects of climate change as represented by the number of hungry people (Figure 2d). The COMP scenario fully offsets the impact of climate change on the population at risk of hunger, driven primarily by investments in R&D in the international agricultural research system, that is, CGIAR. In these scenarios, we assume that investments in national agricultural research systems (NARS) remain at the levels modeled in the reference scenario. While water and infrastructure investments make only small contributions to reducing hunger on their own, they provide a critical foundation of environmental and economic sustainability that underpins longer-term and broader food system resilience.

Underlying the total population at risk of hunger seen in Figure 2 is the share of the population at risk of hunger, which is directly modeled (see Appendix B). While the shares of the population at risk nearly all decline from 2010 to 2050, in a few regions population growth will overwhelm this progress and the projections show increasing numbers of hungry people.

THE COST OF ADAPTATION

Thus far we have described the impact of alternative investments on hunger in the context of socioeconomic and climate change. We are not able to model those impacts directly as functions of dollars invested. Instead, investments are characterized in terms of their intermediate impacts on agricultural productivity (via increases in international agricultural research and improvements in water management) and on marketing costs (via infrastructure improvements). We focus here on R&D in the international agricultural system and set aside investments in NARS, which would require complementary development efforts and increased investments beyond the levels specified in the reference scenario. Next, we turn to estimating the costs of achieving those increases in international agricultural research and improvements in water management and infrastructure.

First, we consider agricultural research. Investment in international agricultural research via CGIAR increased from US$454 million per year in 2005 to almost US$900 million per year in 2015. Assuming in our reference scenario that that historical progression continues as detailed in Rosegrant et al. (2017) and Mason-D'Croz et al. (2019), the level of investment in international agricultural research centers increases by US$1.46 billion from 2015 to 2050, to a total of about US$2.36 billion per year by 2050 (Figure 3a). Using the costing sensitivity analysis explained above, the increase in REF-CC investments in international agricultural research ranges from US$1.30 billion to US$1.69 billion, and the final level in 2050 ranges from US$2.20 billion to US$2.59 billion. The level of investments grows annually by about 2.5–3.0 percent (compounding). The average level of annual investments in the reference scenario from 2015 to 2050 is US$1.72 billion.

The level of R&D investments in the international agricultural research system for the HIGH+RE scenario reaches US$6.68 (US$5.76–8.27) billion per year in 2050. This corresponds to a more than doubling of the growth rate of investments to about 5.7 percent per year, leading to four times the 2015 to 2050 increase seen in the REF-CC scenario (Figure 3b).

The average of BAU investments (the REF-CC line in Figure 3a) from 2015 to 2050 is US$1.72 billion per year (see global total for the CGIAR component of the reference scenario in Table 3), and is the point of comparison with Nelson et al. (2009) (see Appendix A). The average incremental investment level for the increased research investment scenario (HIGH+RE in Figure 3b) from 2015 to 2050 is US$1.97 billion per year (see global total for CGIAR component of the investment scenario in Table 3).

While investments in agricultural R&D go a long way toward offsetting the impact of climate change on hunger, complementary investments in water and infrastructure are necessary for robust, resilient development. This is congruent with Nelson et al. (2009), who argue that the best investments for adapting to

FIGURE 3 PROJECTIONS OF GLOBAL ANNUAL INVESTMENTS NEEDED IN INTERNATIONAL AGRICULTURAL R&D IN THE CGIAR, with minimum and maximum plausible intervals, for the (a) REF-CC and (b) HIGH+RE (incremental to REF) scenarios

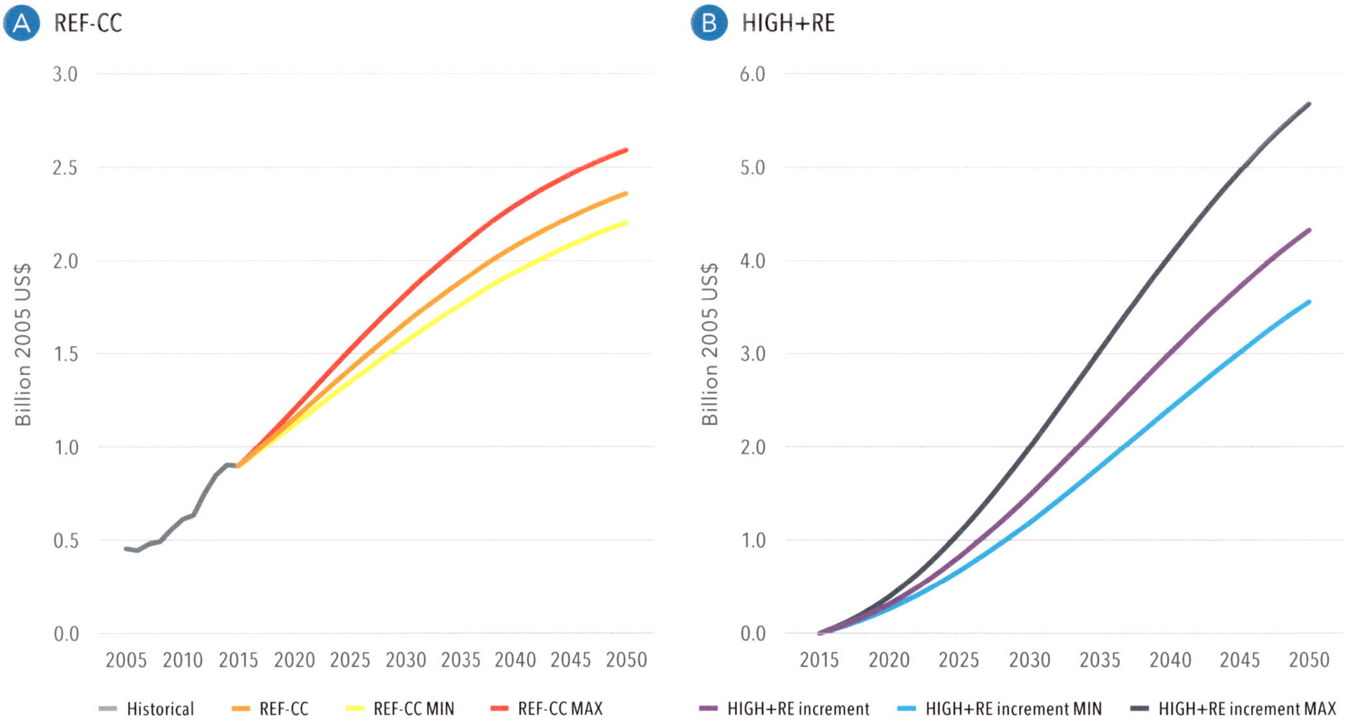

climate change are simply good development investments, which already address resilience.

Table 3 presents the estimates for the cost of R&D, water, and infrastructure investments. Total public investment in these three areas in developing regions is projected to average US$42.56 billion per year between 2015 and 2050 in the reference scenario, with an additional US$25.47 billion per year needed (in the COMP scenario) to offset the impacts of climate change on hunger. As noted above, most of this reduction in hunger is achieved by increased investment in agricultural R&D. Public agricultural R&D spending is projected to average US$8.08 billion per year between 2015 and 2050 in the reference scenario, with an additional US$1.97 billion per year needed to help achieve hunger reduction goals.

Most global agricultural R&D spending, both in the reference case and especially in the investments incremental to reference levels, is focused on SSA. The COMP scenario effectively doubles the CGIAR budget globally (with a range of a 92–148 percent increase); most of the increase is in SSA, where the budget could nearly triple from the reference scenario in the most expensive sensitivity simulation. The CGIAR budget increases the least in EAP (about a 25 percent increase over the reference scenario), while the rest of the world (SAS, MENA, LAC) hovers between a 40 percent to 70 percent increase over the reference scenario. The cost sensitivity simulations for agricultural R&D investments show a range around the reference and COMP scenario cost estimates of +9 percent to −6 percent and +8 percent to −11 percent for the projected value of CGIAR and NARS investments, respectively, in the reference scenario and +41 percent to −24 percent for the projected value of CGIAR investments in the COMP scenario.

Over the longer term, and going beyond the hunger metric, complementary investments in water management and supporting infrastructure are also necessary. Additional water and infrastructure investments are estimated to be more expensive than agricultural R&D, at US$12.7 billion and US$10.8 billion per year, respectively, but they address key gaps that expand over time – such as overused and undermaintained water delivery and drainage systems and inadequate and degrading municipal transport and electrical services – to ensure food system sustainability, efficiency, and reliability. Investments to expand irrigated area and improve water use efficiency and soil water-holding capacity will become increasingly important as climate change makes water resources scarcer in some areas and more erratic in others. Infrastructure investments will be needed to increase the efficiency and reliability of the food system and ensure that rural poor populations are not left behind in areas where the effects of climate change will be felt more severely.

Water and infrastructure investments dominate the cost of the COMP scenario, while infrastructure accounts for much of the reference costs. Reference spending on water investments is mostly in SSA, EAP, and SAS, while incremental investments are concentrated in EAP and SAS. Infrastructure investments are especially focused on EAP, with some in SAS. Infrastructure spending is quite low for SSA in the reference scenario, so increases there will be important.

The sensitivity testing range for water and infrastructure investments is +/–15 percent, which is aligned with the range we see in the costing sensitivity simulations for the agricultural R&D results. Total water investments effectively double from the reference to the investment scenarios, up from a range of US$9.40 to US$12.72 billion per year by an increment of US$10.79 to US$14.59 billion per year in the investment scenario. Infrastructure investments in the COMP scenarios increase by US$9 to US$12 billion over the reference scenario, with most of the

TABLE 3 ANNUAL PUBLIC INVESTMENTS IN THE REFERENCE SCENARIO AND ANNUAL INCREMENTAL COSTS FOR INVESTMENT SCENARIOS TO OFFSET CLIMATE CHANGE EFFECTS ON HUNGER, 2015–2050 (in billion 2005 US$)

Scenarios	East Asia and Pacific	South Asia	Africa south of the Sahara	Middle East and North Africa	Latin America and the Caribbean	Total
Reference scenario						
Agricultural R&D	**1.60 (1.46, 1.63)**	**0.97 (0.89, 1.00)**	**2.23 (1.95, 2.55)**	**1.50 (1.29, 1.68)**	**1.78 (1.68, 1.88)**	**8.08 (7.27, 8.75)**
CGIAR	0.07 (0.06, 0.07)	0.26 (0.26, 0.27)	1.11 (1.03, 1.24)	0.09 (0.08, 0.09)	0.20 (0.19, 0.20)	1.72 (1.62, 1.87)
NARS	1.54 (1.40, 1.56)	0.71 (0.63, 0.73)	1.11 (0.92, 1.31)	1.41 (1.20, 1.59)	1.59 (1.49, 1.68)	6.36 (5.65, 6.88)
Water	**2.57 (2.18, 2.96)**	**2.75 (2.34, 3.16)**	**3.50 (2.98, 4.03)**	**0.99 (0.84, 1.14)**	**1.25 (1.06, 1.44)**	**11.06 (9.4, 12.72)**
IX+WUE	2.23 (1.90, 2.56)	2.58 (2.19, 2.97)	3.11 (2.64, 3.58)	0.88 (0.75, 1.01)	0.97 (0.82, 1.12)	9.77 (8.3, 11.24)
SWHC	0.34 (0.29, 0.39)	0.17 (0.14, 0.20)	0.39 (0.33, 0.45)	0.11 (0.09, 0.13)	0.28 (0.24, 0.32)	1.29 (1.10, 1.48)
Infrastructure	**14.02 (11.92, 16.12)**	**7.54 (6.41, 8.67)**	**0.18 (0.15, 0.21)**	**0.91 (0.77, 1.05)**	**0.77 (0.65, 0.89)**	**23.42 (19.91, 26.93)**
Total	**18.19 (15.56, 20.71)**	**11.26 (9.64, 12.84)**	**5.91 (5.08, 6.79)**	**3.40 (2.90, 3.86)**	**3.80 (3.40, 4.20)**	**42.56 (36.58, 48.40)**
Investment scenarios (incremental to reference scenario)						
Agricultural R&D	**0.02 (0.01, 0.02)**	**0.14 (0.12, 0.17)**	**1.66 (1.22, 2.40)**	**0.04 (0.03, 0.05)**	**0.11 (0.10, 0.13)**	**1.97 (1.49, 2.77)**
CGIAR	0.02 (0.01, 0.02)	0.14 (0.12, 0.17)	1.66 (1.22, 2.40)	0.04 (0.03, 0.05)	0.11 (0.10, 0.13)	1.97 (1.49, 2.77)
Water	**3.54 (3.01, 4.07)**	**2.60 (2.21, 2.99)**	**2.62 (2.23, 3.01)**	**1.53 (1.30, 1.76)**	**2.40 (2.04, 2.76)**	**12.69 (10.79, 14.59)**
IX+WUE	2.86 (2.43, 3.29)	1.65 (1.40, 1.90)	1.42 (1.21, 1.63)	0.65 (0.55, 0.75)	1.53 (1.30, 1.76)	8.11 (6.89, 9.33)
SWHC	0.68 (0.58, 0.78)	0.95 (0.81, 1.09)	1.20 (1.02, 1.38)	0.88 (0.75, 1.01)	0.87 (0.74, 1.00)	4.58 (3.89, 5.27)
Infrastructure	**3.81 (3.24, 4.38)**	**2.20 (1.87, 2.53)**	**1.90 (1.62, 2.19)**	**0.86 (0.73, 0.99)**	**2.04 (1.73, 2.35)**	**10.81 (9.19, 12.43)**
Comprehensive	**7.37 (6.26, 8.47)**	**4.94 (4.20, 5.69)**	**6.18 (5.07, 7.60)**	**2.43 (2.07, 2.80)**	**4.55 (3.87, 5.24)**	**25.47 (21.47, 29.80)**

Note: Regions only include developing countries. Data shown are a reference estimate with the minimum and maximum of a plausible interval (in parentheses) using methods described.

TABLE 4 SDG TRADE-OFF TABLES: SCENARIO IMPACTS ON SELECTED OUTCOME INDICATORS IN 2030 AND 2050
(costs are in billion US$ per year for the developing world; other values are percentage differences relative to reference scenario)

		2030						
		SDG 1	SDG 2			SDG 6/12		
Scenario	Average Annual Cost	GDP	Ag Supply	Hunger	Per Capita KCAL Availability	Blue Water Use	Irrigation Water Reliability	Blue Water per KCAL
HIGH+RE	1.97	1.73	6.29	-24.64	4.19	0.21	0.40	-4.22
IX+WUE	8.11	0.23	0.87	-4.28	0.73	-7.04	12.88	-7.71
SWHC	4.58	0.14	0.52	-2.17	0.31	-1.48	0.98	-1.78
RMM	10.81	0.37	0.08	-0.44	0.06	-0.02	-0.03	-0.08
COMP	25.47	2.38	7.34	-28.16	5.00	-7.41	13.09	-11.82

		2050						
		SDG 1	SDG 2			SDG 6/12		
Scenario	Average Annual Cost	GDP	Ag Supply	Hunger	Per Capita KCAL Availability	Blue Water Use	Irrigation Water Reliability	Blue Water per KCAL
HIGH+RE	1.97	3.43	7.45	-21.55	5.63	-0.41	0.37	-5.71
IX+WUE	8.11	0.28	0.83	-3.29	0.70	-7.39	12.13	-8.04
SWHC	4.58	0.46	0.91	-3.77	0.71	-2.88	2.02	-3.57
RMM	10.81	0.54	0.14	-0.24	0.08	-0.04	-0.06	-0.12
COMP	25.47	4.62	8.62	-23.66	6.50	-7.95	12.30	-13.57

Heatmap Legend — Less Advantageous — Neutral — More Advantageous

spending in EAP and SAS, though SSA sees the largest increase (by a factor of 10) from a smaller starting point.

TRADE-OFFS

The alternative investment scenarios entail a wide range of costs and generate a wide range of outcomes for the SDGs. In some cases, the impacts on different objectives are complementary; in others, significant trade-offs arise between objectives.

Table 4 summarizes key impacts of each scenario across SDGs compared to the REF-CC scenario. Investments in agricultural R&D in the HIGH+RE scenario are relatively low cost and generally offer moderate improvements in income by 2030, but larger improvements by 2050. The HIGH+RE investments make important contributions to increasing food availability and reducing hunger, while also improving the water situation around the globe. The scenario combining irrigation expansion and increased water use efficiency (IX+WUE) offers greater reductions in water use (with improved efficiency) at a greater cost, but comes with smaller improvements in income, food supply, and food security. Improved market access through reduced marketing costs (RMM) increases income, supply, and food security, but at the cost of water with no improvements to use and decreased irrigation water reliability. These outcomes highlight the importance of a mixed portfolio of investments that combines productivity enhancement with improved water

TABLE 5 PLAUSIBLE RANGES FOR SDG TRADE-OFF TABLES; SCENARIO IMPACTS ON SELECTED OUTCOME INDICATORS IN 2030 AND 2050
(costs are in billion US$ per year for the developing world; other values are percentage differences relative to reference scenario)

		2030						
		SDG 1	SDG 2			SDG 6/12		
Scenario	Average Annual Cost	GDP	Ag Supply	Hunger	Per Capita KCAL Availability	Blue Water Use	Irrigation Water Reliability	Blue Water per KCAL
HIGH+RE	(1.49, 2.77)	(1.69, 11.16)	(6.28, 7.08)	(-34.56, -22.10)	(4.04, 6.90)	(-0.70, -0.20)	(-0.10, 0.42)	(-4.26, -3.37)
IX+WUE	(6.89, 9.33)	(-0.17, 9.64)	(0.77, 1.55)	(-23.98, -3.82)	(0.66, 3.66)	(-7.39, -6.91)	(12.69, 13.02)	(-8.00, -6.93)
SWHC	(3.89 ,5.27)	(0.10, 9.52)	(0.51, 1.09)	(-22.53, -2.12)	(0.31, 3.15)	(-1.98, -1.47)	(0.50, 0.98)	(-1.79, -1.15)
RMM	(9.19, 12.43)	(0.34, 9.76)	(0.07, 0.65)	(-20.77, -0.41)	(0.06, 2.90)	(-0.55, -0.01)	(-0.52, -0.03)	(-0.08, 0.53)
COMP	(21.47, 29.80)	(2.33, 11.52)	(7.34, 8.16)	(-36.03, -24.00)	(4.84, 7.74)	(-7.80, -7.32)	(12.82, 13.16)	(-12.06, -10.74)

		2050						
		SDG 1	SDG 2			SDG 6/12		
Scenario	Average Annual Cost	GDP	Ag Supply	Hunger	Per Capita KCAL Availability	Blue Water Use	Irrigation Water Reliability	Blue Water per KCAL
HIGH+RE	(1.49, 2.77)	(3.14, 33.73)	(7.34, 11.07)	(-44.11, -15.36)	(5.05, 13.5)	(-2.84, -0.34)	(-2.42, 0.47)	(-5.71, -5.21)
IX+WUE	(6.89, 9.33)	(-0.86, 29.92)	(0.53, 3.93)	(-34.13, -2.04)	(0.48, 9.23)	(-9.34, -6.93)	(9.78, 12.64)	(-8.47, -7.58)
SWHC	(3.89 ,5.27)	(0.42, 30.14)	(0.90, 3.88)	(-35.42, -3.45)	(0.71, 9.15)	(-5.35, -2.86)	(-0.65, 2.03)	(-4.38, -3.55)
RMM	(9.19, 12.43)	(0.47, 30.20)	(0.12, 3.11)	(-32.42, -0.11)	(0.07, 8.51)	(-2.50, -0.03)	(-2.72, -0.06)	(-0.92, -0.10)
COMP	(21.47, 29.80)	(4.15, 34.60)	(8.53, 12.25)	(-44.93, -16.29)	(5.89, 14.39)	(-9.75, -7.60)	(9.98, 12.59)	(-13.74, -13.00)

Heatmap Legend
Less Advantageous — Neutral — More Advantageous

resource management and market access. The comprehensive scenario (COMP) achieves significant improvements in all outcome areas, particularly in 2050, but comes at a significantly higher cost.

The plausible ranges of scenario impacts (Table 5) also include important details to consider. The potential income gains across all the investment scenarios are quite large, with around a 10 percent increase by 2030 and more than a 30 percent increase by 2050; however, at the lower end of this range we see rather modest income improvements to even slightly negative effects with IX+WUE. The HIGH+RE carries with it important improvements for SDG 2 even at the lowest end of the plausible range. The other scenarios have a lot of potential to improve the hunger situation, but also show that these investments might not result in strong reductions in hunger. Water use and efficiency (SDGs 6 and 12) is clearly improved under the IX+WUE investment scenario, but HIGH+RE makes important contributions in this realm, too. The other particular scenarios (SWHC and RMM) show mixed improvements to water use and efficiency.

Beyond the SDGs, results are varied for the different commodities for regions across the different scenarios, which may be of interest to policymakers (Appendix C).

DISCUSSION AND CONCLUSIONS

KEY DIFFERENCES FROM OTHER STUDIES

Nelson et al. (2009) previously analyzed the cost of adapting to climate change by increasing investments in agriculture and food systems in the developing world. They estimated those additional annual costs to be around US$1 billion for agricultural research and US$3 billion each for irrigation and rural roads. The current analysis builds on the same general approach but generates corresponding estimates roughly 2–4 times higher: around US$2 billion for agricultural R&D, US$13 billion for water management, and US$11 billion for rural infrastructure. These differences arise for several key reasons:

- Nelson et al. (2009) analyzed a scenario designed to offset the effects of climate change as measured by the number of undernourished children. The current analysis, building on Rosegrant et al. (2017), analyzes a range of plausible investment scenarios developed in collaboration with scientists from across the CGIAR and estimates their costs and impacts, including on the number of people facing chronic hunger.

- The current analysis uses an updated modeling specification in conjunction with a variety of updated modeling assumptions, including updated versions of the IMPACT model parameter specifications (for supply/demand elasticities, embedded technological progress assumptions, and geography specifications, among others), and related modeling methods within the IMPACT framework (crop and water models, in particular).

- The current analysis uses more recent climate[2] and socioeconomic data[3] based on the IPCC's Fifth Assessment Report (IPCC 2014).

- The current analysis uses an updated methodology based on the perpetual inventory method to capture the lagged effects of R&D investment on productivity.

As noted at the beginning of this report, earlier studies estimated the cost of adaptation to climate change at US$75–100 billion. This is considerably higher than our estimates, but we emphasize that our analysis looks only at the cost of adaptation in a single dimension – in terms of offsetting the adverse impacts of climate change on hunger.

Other studies have explored investments to reduce hunger, particularly in Africa south of the Sahara. These include studies by Laborde et al. (2016) and FAO, IFAD, and WFP (2015). Direct comparison is difficult because these other analyses focused on different aspects of the question, used different methods, and did not address the effects of climate change. Laborde et al. (2016) considered the minimum cost to eliminate hunger in vulnerable households; FAO, IFAD, and WFP (2015) assessed the cost to lift all households above the poverty line assuming it will be equivalent to a bulk cash trans-

---

2 The temperature and precipitation for the IPCC AR4 climates used in the Nelson et al. (2009) study differ from the various IPCC AR5 RCP and GCM combinations for the current study in many ways across the planet's surface, which leads to different modeling outcomes.

3 Socioeconomic assumptions to 2050 used by Nelson et al. (2009) were derived from various United Nations and World Bank sources, while the SSP assumptions used in the present analysis were developed in a more systematic way as part of the IPCC AR5 process. For example, per capita GDP globally for the currently used SSP2 in 2050 is US$25,000, whereas in Nelson et al. (2009) it was US$17,000. This has a strong impact on results.

fer. Both studies focused on a reference case that is equivalent to our REF-NoCC, while our analysis specifically focuses on climate change effects. Other key differences between the various analyses are summarized in Fan et al. (2018) and Mason-D'Croz et al. (2019).

CAVEATS, LIMITATIONS, AND FUTURE ADVANCES

While Rosegrant et al. (2017) made important advances in the estimation of costs and impacts of investments for agriculture, it is important to recognize several limitations of the current study and the opportunities to further advance the analysis of agriculture and food system investments for climate change adaptation.

The methodology underlying this analysis is still evolving. Both specific algorithms and parameterization of various factors are active fields of study and will likely see improvements in the future.

The results arrived at here might underestimate full costs. For example, trade plays an important role in adaptation in these analyses but is modeled in a relatively frictionless way in the IMPACT model framework (Janssens et al. 2020). Trade barriers will increase adaptation costs, especially for net importers. Further work is needed to better reflect the reality of evolving trade policies and trade interrelationships.

While the IMPACT modeling framework used here included feedback from a full economy model on total GDP, modeling improvements are needed to more fully integrate feedback from other sectors of the economy (energy, labor, etc.), private sector investment, and other complementary development strategies – such as social safety nets – for any particular climate scenario. These dimensions could either increase or decrease the total cost of adaptation; a fully integrated model is required to determine the direction and magnitude of change attributable to these different factors.

This analysis focuses on projected changes in climate trends (specifically temperature and precipitation) but does not capture the effects of changes in climate variability or extreme events (or other aspects of climate change such as sea level rise or melting glaciers).

This analysis focuses on just one dimension and indicator of adaptation to climate change: the number of people facing chronic hunger. Though we added breadth to this in our additional SDG tradeoff assessment, other SDG targets could be assessed with further development of modeling tools that can explicitly address the multitude of SDG dimensions.[4] A full analysis of adaptation to climate change in agriculture would require including many other social, economic, and environmental dimensions. Several dimensions exist in which to consider tradeoffs across the range of SDGs in any set of scenarios of future shocks and adaptation investments. In addition to the trade-offs presented here, results on several of these additional indicators and their tradeoffs are reported in Rosegrant et al. (2017). There are also many additional pathways of climate change impact on food systems and human well-being (Mbow et al. 2019; Myers et al. 2017; Crimmins et al. 2016), including altered production systems, vector-borne diseases, water source quality, and food safety, among many others.

Finally, this study is not designed to identify the minimum cost of adaptation to climate change. Rather, it examines several alternative investment strategies, their costs, and their impacts in relation to the selected indicator of adaptation. Further analysis

4 We note that SDG 13 on "Climate Action, Take urgent action to combat climate change and its impacts" has a special relation to this analysis. Indeed, the investment alternatives modeled here are actions that can be taken to increase adaptive capacity to a changing climate and reduce susceptibility to climate disasters. These types of actions could be embedded in policy and planning processes at the national and regional level. This study is focused on adaptation, but, as shown in Rosegrant et al. (2017), there are important mitigation benefits to several of these investment alternatives, with reduced land use and greenhouse gas emissions by the agriculture sector. The investment scenarios are set up as assumptions for assessing alternatives, however, and are not results drawn from the modeling framework, although co-benefits arise, as noted. For this reason, we did not include SDG 13 as an integral part of this analysis.

is required to provide a more detailed picture of the range of possible options, including a broader set of investment alternatives and other policy approaches that aim at not only supply-side levers but also cost-effective ways to affect levels of demand and diet preferences, food waste, and food system "greening," among others, which could ease food system stressors in the face of climate change.

KEY MESSAGES

Planning for an uncertain climate future will require an enabling environment for innovation in agricultural technologies and practices that can be applied across broad geographies and adapted to local requirements. Increased investment and innovation in agriculture need to be complemented by robust rural and pro-poor development strategies. The analysis presented here offers some key messages about the costs and trade-offs of investing in R&D, water management, and infrastructure for climate-change adaptation to help reach the goal of zero hunger:

- Climate change is projected to slow progress in reducing hunger to 2050, particularly in Africa south of the Sahara.

- Increased investment in agriculture can support adaptation to climate change. R&D is most cost-effective in offsetting the impacts of climate change as measured by the hunger indicator, with US$1.97 billion in annual incremental investment needed for adaptation by this measure.

- Estimating ranges of results gives some context about the potential for investments to either under- or overshoot highlighted impacts, along with a better idea about costs that should be considered as part of the trade-offs.

- Increased investment in water management and infrastructure is essential to complement and sustain R&D-based gains, especially over the longer term. Other investments will be needed to achieve broader measures of adaptation.

- Alternative investment options involve different synergies and trade-offs: R&D investments offer the largest reductions in hunger but smaller improvements in blue water use and irrigation supply reliability; water management investments offer greater improvements in blue water use and irrigation supply but smaller reductions in hunger; and a comprehensive approach offers improvements in all goals but is much more costly.

REFERENCES

Agrawala, S., and S. Fankhauser, eds. 2008. *Economic Aspects of Adaptation to Climate Change: Costs, Benefits and Policy Instruments*. Paris: OECD Publishing. https://doi.org/10.1787/9789264046214-en

Baldos, U.L.C., K.O. Fuglie, and T.W. Hertel. 2020. "The Research Cost of Adapting Agriculture to Climate Change: A Global Analysis to 2050." *Agricultural Economics* 51: 207–220. https://doi.org/10.1111/agec.12550

Beach, R.H., T.B. Sulser, A. Crimmins, N. Cenacchi, J. Cole, N.K. Fukagawa, D. Mason-D'Croz, S. Myers, M.C. Sarofim, M. Smith, and L.H. Ziska. 2019. "Combining the Effects of Increased Atmospheric Carbon Dioxide on Protein, Iron, and Zinc Availability and Projected Climate Change on Global Diets: A Modelling Study." *Lancet Planetary Health* 3: e307-e317. https://doi.org/10.1016/S2542-5196(19)30094-4

Chambwera, M., G. Heal, C. Dubeux, S. Hallegatte, L. Leclerc, A. Markandya, B.A. McCarl, R. Mechler, and J.E. Neumann. 2014. "Economics of Adaptation." In *Climate Change 2014: Impacts, Adaptation, and Vulnerability. Part A: Global and Sectoral Aspects. Contribution of Working Group II to the Fifth Assessment Report of the Intergovernmental Panel on Climate Change*, ed. C.B. Field, et al., chapter 17. Cambridge: Cambridge University Press. https://www.ipcc.ch/report/ar5/wg2/

Chan, C.Y., N. Tran, S. Pethiyagoda, C.C. Crissman, T.B. Sulser, and M.J. Phillips. 2019. "Prospects and Challenges of Fish for Food Security in Africa." *Global Food Security* 20: 17–25. https://doi.org/10.1016/j.gfs.2018.12.002

Crimmins, A., J. Balbus, J.L. Gamble, C.B. Beard, J.E. Bell, D. Dodgen, R.J. Eisen, N. Fann, M.D. Hawkins, S.C. Herring, L. Jantarasami, D.M. Mills, S. Saha, M.C. Sarofim, J. Trtanj, and L. Ziska, eds. 2016. *The Impacts of Climate Change on Human Health in the United States: A Scientific Assessment*. Washington, DC: U.S. Global Change Research Program. https://doi.org/10.7930/J0R49NQX

Enahoro, D., D. Mason-D'Croz, M. Mul, K.M. Rich, T.P. Robinson, P. Thornton, and S.S. Staal. 2019. "Supporting Sustainable Expansion of Livestock Production in South Asia and Sub-Saharan Africa: Scenario Analysis of Investment Options." *Global Food Security* 20: 114-121. https://doi.org/10.1016/j.gfs.2019.01.001

Esposti, M., and F. Pierani. 2003. "Building the Knowledge Stock: Lags, Depreciation, and Uncertainty in R&D Investment and Link with Productivity Growth." *Journal of Productivity Analysis* 19: 33-58. https://doi.org/10.1023/A:1021818019626

Fan, S., D.D. Headey, D. Laborde, D. Mason-D'Croz, C. Rue, T.B. Sulser, and K. Wiebe. 2018. *Quantifying the Cost and Benefits of Ending Hunger and Undernutrition: Examining the Differences Among Alternative Approaches*. IFPRI Issue Brief. Washington, DC: IFPRI. https://doi.org/10.2499/9780896292994

Fankhauser, S. 2010. "The Costs of Adaptation." *WIREs Climate Change* 1: 23–30. https://doi.org/10.1002/wcc.14

FAO (Food and Agriculture Organization of the United Nations). 2008. "FAO Methodology for the Measurement of Food Deprivation Updating the Minimum Dietary Energy Requirements." Rome. http://www.fao.org/fileadmin/templates/ess/documents/food_security_statistics/metadata/undernourishment_methodology.pdf

——. 2016. *AQUASTAT Database on Investment Costs in Irrigation*. Rome. http://www.fao.org/nr/water/aquastat/investment/index.stm

FAO, International Fund for Agricultural Development (IFAD), and World Food Programme (WFP). 2015. *Achieving Zero Hunger: The Critical Role of Investments in Social Protection and Agriculture*. Rome: FAO. http://www.fao.org/3/a-i4951e.pdf

Forstater, M., and S. Zadek. 2009. "Low Carbon Growth Plans: Assessing Current Practice." London: Project Catalyst.

Frija, A., A. Chebil, K.A. Mottaleb, D. Mason-D'Croz, and B. Dhehibi. 2020. "Agricultural Growth and Sex-Disaggregated Employment in Africa: Future Perspectives under Different Investment Scenarios." *Global Food Security* 24: 100353. https://doi.org/10.1016/j.gfs.2020.100353

Inocencio, A., M. Kikuchi, M. Tonosaki, A. Maruyama, D. Merrey, H. Sally, and I. de Jong. 2007. *Cost of Performance of Irrigation Projects: A Comparison of Sub-Saharan Africa and Other Developing Regions*. IWMI Research Report 109. Colombo: International Water Management Institute. https://doi.org/10.3910/2009.109

IPCC (Intergovernmental Panel on Climate Change). 2014. *Climate Change 2014: Impacts, Adaptation, and Vulnerability. Part A: Global and Sectoral Aspects. Contribution of Working Group II to the Fifth Assessment Report of the Intergovernmental Panel on Climate Change*, ed.C.B. Field et al. Cambridge: Cambridge University Press. https://www.ipcc.ch/report/ar5/wg2/

IWMI (International Water Management Institute). 2012. *AgWater Solutions Database*. Colombo. http://agwaterdb.iwmi.org/

Janssens, C., P. Havlík, T. Krisztin, J. Baker, S. Frank, T. Hasegawa, D. Leclère, D. Ohrel, S. Ragnauth, E. Schmid, H. Valin, N. Van Lipzig, and M. Maertens. 2020. "Global Hunger and Climate Change Adaptation Through International Trade." *Nature Climate Change* 10: 829–835. https://doi.org/10.1038/s41558-020-0847-4

Kruseman, G., K.A. Mottaleb, K. Tesfaye, S. Bairagi, R. Robertson, D. Mandiaye, A. Frija, S. Gbegbelegbe, A. Alene, and S. Prager. 2020. "Rural Transformation and the Future of Cereal-Based Agri-Food Systems." *Global Food Security* 26: 100441. https://doi.org/10.1016/j.gfs.2020.100441

Laborde, D., L. Bizikova, T. Lallemant, and C. Smaller. 2016. *Ending Hunger: What Would It Cost?* Winnipeg, Canada: International Institute for Sustainable Development. https://www.iisd.org/publications/ending-hunger-what-would-it-cost

Mason-D'Croz, D., T.B. Sulser, K. Wiebe, M.W. Rosegrant, S.K. Lowder, A. Nin-Pratt, D. Willenbockel, S. Robinson, T. Zhu, N. Cenacchi, S. Dunston, and R.D. Robertson. 2019. "Agricultural Investments and Hunger in Africa Modeling Potential Contributions to SDG2 – Zero Hunger." *World Development* 116: 38–53. https://doi.org/10.1016/j.worlddev.2018.12.006

Mbow, C., C. Rosenzweig, L.G. Barioni, T.G. Benton, M. Herrero, M. Krishnapillai, E. Liwenga, P. Pradhan, M.G. Rivera-Ferre, T. Sapkota, F.N. Tubiello, and Y. Xu. 2019. "Food Security." In *Climate Change and Land: An IPCC Special Report on Climate Change, Desertification, Land Degradation, Sustainable Land Management, Food Security, and Greenhouse Gas Fluxes in Terrestrial Ecosystems*, ed. P.R. Shukla et al., Chapter 5. Geneva: Intergovernmental Panel on Climate Change.

Moss, R.H., J.A. Edmonds, K.A. Hibbard, M.R. Manning, S.K. Rose, D.P. van Vuuren, T.R. Carter, S. Emori, M. Kainuma, T. Kram, G.A. Meehl, J.F.B. Mitchell, N. Nakicenovic, K. Riahi, S.J. Smith, R.J. Stouffer, A.M. Thomson, J.P. Weyant, and T.J. Wilbanks. 2010. "The Next Generation of Scenarios for Climate Change Research and Assessment." *Nature* 463: 747–756. https://doi.org/10.1038/nature08823

Myers, S.S., M.R. Smith, S. Guth, C.D. Golden, B. Vaitla, N.D. Mueller, A.D. Dangour, and P. Huybers. 2017. "Climate Change and Global Food Systems: Potential Impacts on Food Security and Undernutrition." *Annual Review of Public Health* 38: 259–277. https://doi.org/10.1146/annurev-publhealth-031816-044356

Navarro-Racines, C., J. Tarapues, P. Thornton, A. Jarvis, and J. Ramirez-Villegas. 2020. "High-Resolution and Bias-Corrected CMIP5 Projections for Climate Change Impact Assessments." *Scientific Data* 7: 7. https://doi.org/10.1038/s41597-019-0343-8

Nelson, G., J. Bogard, K. Lividini, J. Arsenault, M. Riley, T.B. Sulser, D. Mason-D'Croz, B. Power, D. Gustafson, M. Herrero, K. Wiebe, K. Cooper, R. Remans, and M. Rosegrant. 2018. "Income Growth and Climate Change Effects on Global Nutrition Security to Mid-Century." *Nature Sustainability* 1: 773–781. https://doi.org/10.1038/s41893-018-0192-z

Nelson, G.C., M.W. Rosegrant, J. Koo, R. Robertson, T.B. Sulser, T. Zhu, C. Ringler, S. Msangi, A. Palazzo, M. Batka, M. Magalhaes, R. Valmonte-Santos, M. Ewing, and D. Lee. 2009. *Climate Change: Impact on Agriculture and Costs of Adaptation*. Food Policy Report No. 21. Washington, DC: IFPRI. https://doi.org/10.2499/0896295354

O'Garra, T., and S. Mourato. 2016. "Are We Willing to Give What It Takes? Willingness to Pay for Climate Change Adaptation in Developing Countries." *Journal of Environmental Economics and Policy* 5: 249–264. https://doi.org/10.1080/21606544.2015.1100560

O'Neill, B.C., E. Kriegler, K.L. Ebi, E. Kemp-Benedict, K. Riahi, D.S. Rothman, B.J. van Ruijven, D.P. van Vuuren, J. Birkmann, K. Kok, M. Levy, and W. Solecki. 2017. "The Roads Ahead: Narratives for Shared Socioeconomic Pathways Describing World Futures in the 21st Century." *Global Environmental Change* 42: 169–180. https://doi.org/10.1016/j.gloenvcha.2015.01.004

Petsakos, A., S.D. Prager, C.E. Gonzalez, A.C. Gama, T.B. Sulser, S. Gbegbelegbe, E.M. Kikulwe, and G. Hareau. 2019. "Understanding the Consequences of Changes in the Production Frontiers for Roots, Tubers and Bananas." *Global Food Security* 20: 180–188. https://doi.org/10.1016/j.gfs.2018.12.005

Robinson, S., D. Mason-D'Croz, S. Islam, T.B. Sulser, R.D. Robertson, T. Zhu, A. Gueneau, G. Pitois, and M.W. Rosegrant. 2015. *The International Model for Policy Analysis of Agricultural Commodities and Trade (IMPACT): Model Description for Version 3*. IFPRI Discussion Paper 1483, IFPRI, Washington, DC. https://ebrary.ifpri.org/digital/collection/p15738coll2/id/129825

Rosegrant, M.W., E. Magalhaes, R. Valmonte-Santos, and D. Mason-D'Croz. 2018. "Returns to Investment in Reducing Postharvest Food Losses and Increasing Agricultural Productivity Growth." In *Prioritizing Development: A Cost Benefit Analysis of the United Nations' Sustainable Development Goals*, ed. B. Lomborg. Cambridge: Cambridge University Press. https://doi.org/10.1017/9781108233767.020

Rosegrant, M.W., T.B. Sulser, D. Mason-D'Croz, N. Cenacchi, A. Nin-Pratt, S. Dunston, T. Zhu, C. Ringler, K. Wiebe, S. Robinson, D. Willenbockel, H. Xie, H.Y. Kwon, T. Johnson, T.S. Thomas, F. Wimmer, R. Schaldach, G.C. Nelson, and B. Willaarts. 2017. *Quantitative Foresight Modeling to Inform the CGIAR Research Portfolio*. Project Report for USAID. Washington, DC: IFPRI. https://ebrary.ifpri.org/digital/collection/p15738coll2/id/131144

Rosegrant, M.W., K.D. Wiebe, T.B. Sulser, D. Mason-D'Croz, and D. Willenbockel. 2021. "Climate Change and Agricultural Development." In *Agricultural Development: New Perspectives in a Changing World*, ed. K. Otsuka, and S. Fan, chapter 19. Washington, DC: IFPRI. https://doi.org/10.2499/9780896293830_19

Tol, R.S.J. 2005. "The Marginal Damage Costs of Carbon Dioxide Emissions: An Assessment of the Uncertainties." *Energy Policy* 33: 2064–2074. https://doi.org/https://doi.org/10.1016/j.enpol.2004.04.002

Tol, R.S.J., S. Fankhauser, and J.B. Smith. 1998. "The Scope for Adaptation to Climate Change: What Can We Learn from the Impact Literature?" *Global Environmental Change* 8: 109–123. https://doi.org/https://doi.org/10.1016/S0959-3780(98)00004-1

UN (United Nations). 2015. *Transforming Our World: The 2030 Agenda for Sustainable Development*. A/RES/70/1. New York: United Nations. https://sustainabledevelopment.un.org/content/documents/21252030%20Agenda%20for%20Sustainable%20Development%20web.pdf

UNFCC (United Nations Framework Convention on Climate Change). 2007. *Investment and Financial Flows to Address Climate Change*. Bonn: UNFCC. https://unfccc.int/files/essential_background/background_publications_htmlpdf/application/pdf/pub_07_financial_flows.pdf

World Bank. 2006. *Investment Framework for Clean Energy and Development*. Washington, DC: World Bank. https://documents.worldbank.org/curated/en/338911468157194334/An-investment-framework-for-clean-energy-and-develop ment-an-investment-framework-clean-energy-and-development

World Bank. 2010. *Economics of Climate Change: Synthesis Report*. Washington, DC: World Bank. https://documents.worldbank.org/curated/en/646291468171244256/Economics-of-adaptatio n-to-climate-change-Synthesis-report

ABOUT THE AUTHORS

Timothy B. Sulser is a senior scientist, **Keith Wiebe** is a senior research fellow, **Shahnila Dunston** and **Nicola Cenacchi** are senior research analysts, **Alejandro Nin-Pratt** is a senior research fellow, and **Richard Robertson** is a research fellow, all with the Environment and Production Technology Division (EPTD), International Food Policy Research Institute (IFPRI), Washington, DC. **Mark W. Rosegrant** is a former division director and research fellow emeritus, IFPRI, Washington, DC. **Daniel Mason-D'Croz** is a senior research scientist, Commonwealth Scientific and Industrial Research Organisation, St Lucia, Australia. **Dirk Willenbockel** is a research fellow, Institute of Development Studies, University of Sussex, Brighton, UK.

SUPPLEMENTARY MATERIALS

Climate Change and Hunger: Estimating Costs of Adaptation in the Agrifood System

APPENDIXES A – D

Appendix A. Details of estimates from Nelson et al. (2009)

Table A – 1 Reproduction of Table 10 in Nelson et al. 2009 ("Additional annual investment expenditure needed to counteract the effects of climate change on nutrition (million 2000 US$)")

	South Asia	East Asia and the Pacific	Europe and Central Asia	Latin America and the Caribbean	Middle East and North Africa	Sub-Saharan Africa	Developing Countries
National Center for Atmospheric Research model with developing country investments							
Agricultural Research	172	151	84	426	169	314	1316
Irrigation Expansion	344	15	6	31	-26	537	907
Irrigation Efficiency	999	686	99	129	59	187	2159
Rural Roads (area)	8	73	0	573	37	1980	2671
Rural Roads (yield)	9	9	10	3	1	35	67
Total	1532	934	199	1162	240	3053	7120
Commonwealth Scientific and Industrial Research Organisation model with developing country investments							
Agricultural Research	185	172	110	392	190	326	1375
Irrigation Expansion	344	1	1	30	-22	529	883
Irrigation Efficiency	1006	648	101	128	58	186	2127
Rural Roads (area)	16	147	0	763	44	1911	2881
Rural Roads (yield)	13	9	11	3	1	36	73
Total	1564	977	223	1316	271	2988	7339

Source: Nelson et al 2009.

Appendix B. Share of population at risk of chronic hunger

Figure B – 1 Share of population at risk of hunger from 2010 to 2050 across ranges of core driver assumptions in the IMPACT model

a) Across SSPs under NoCC from 2010 to 2050

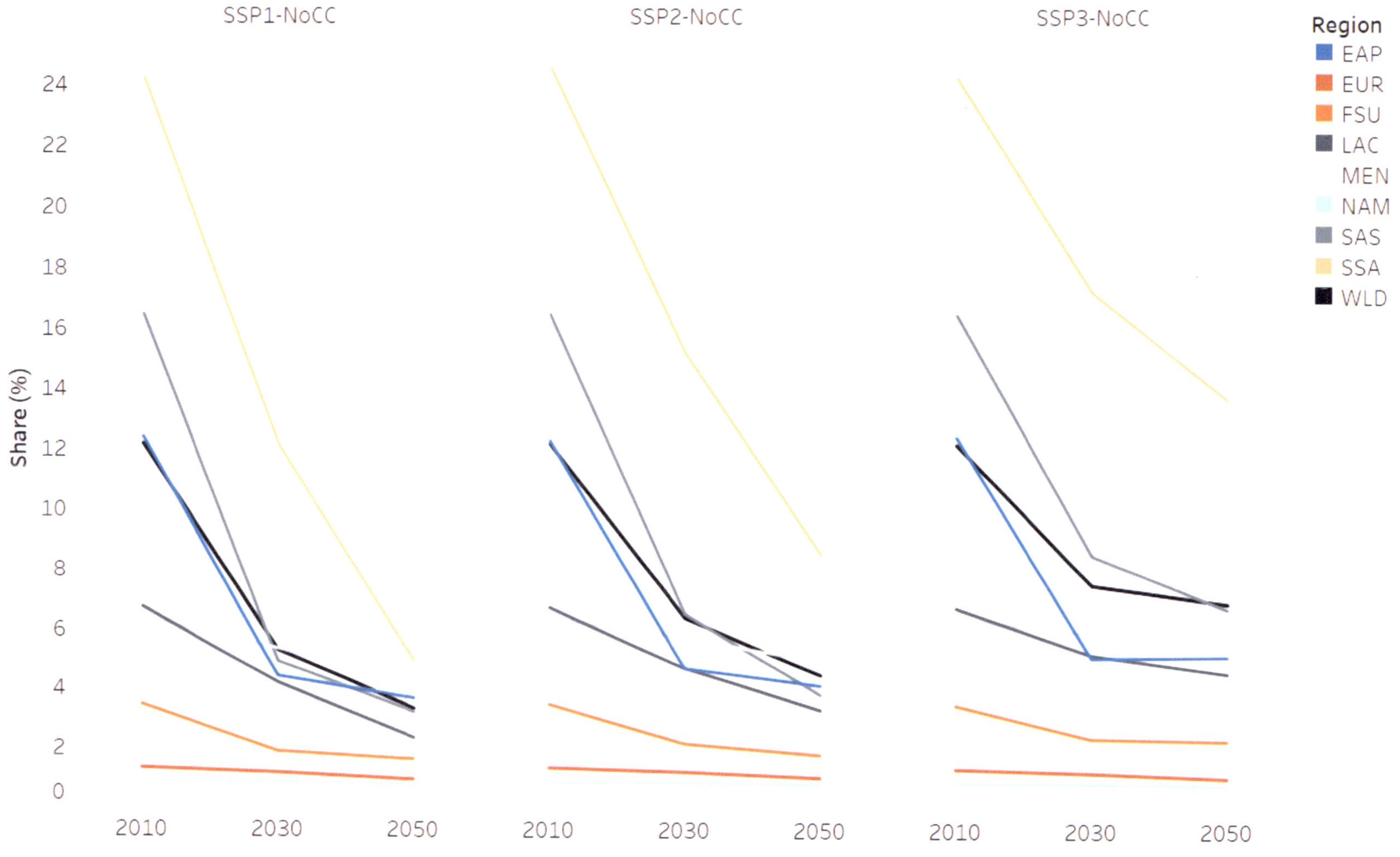

b) Across GCMs for RCP8.5 under SSP2 in 2050

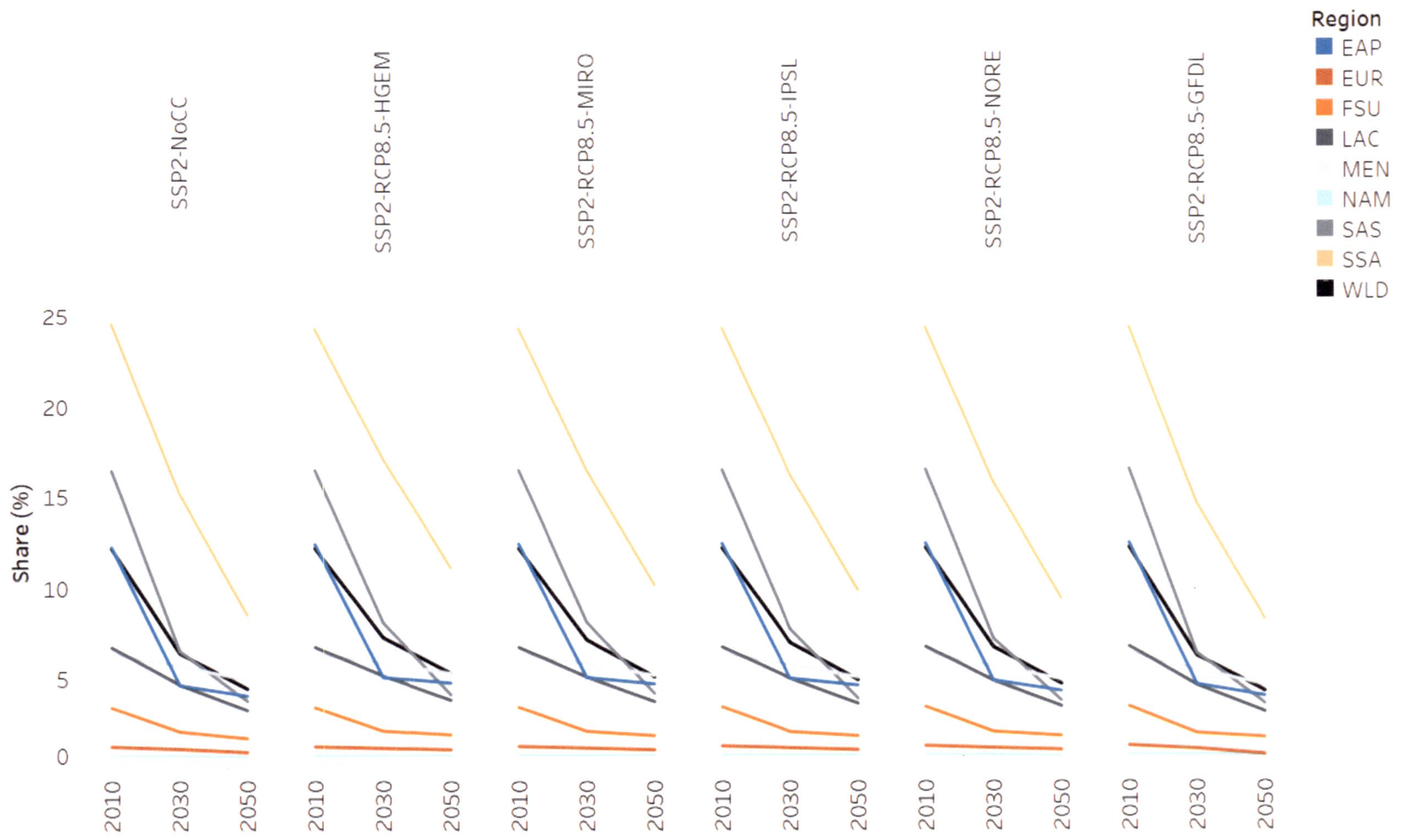

c) Across SSP scenarios combined with RCP4.5 and RCP8.5 with HGEM GCM in 2050

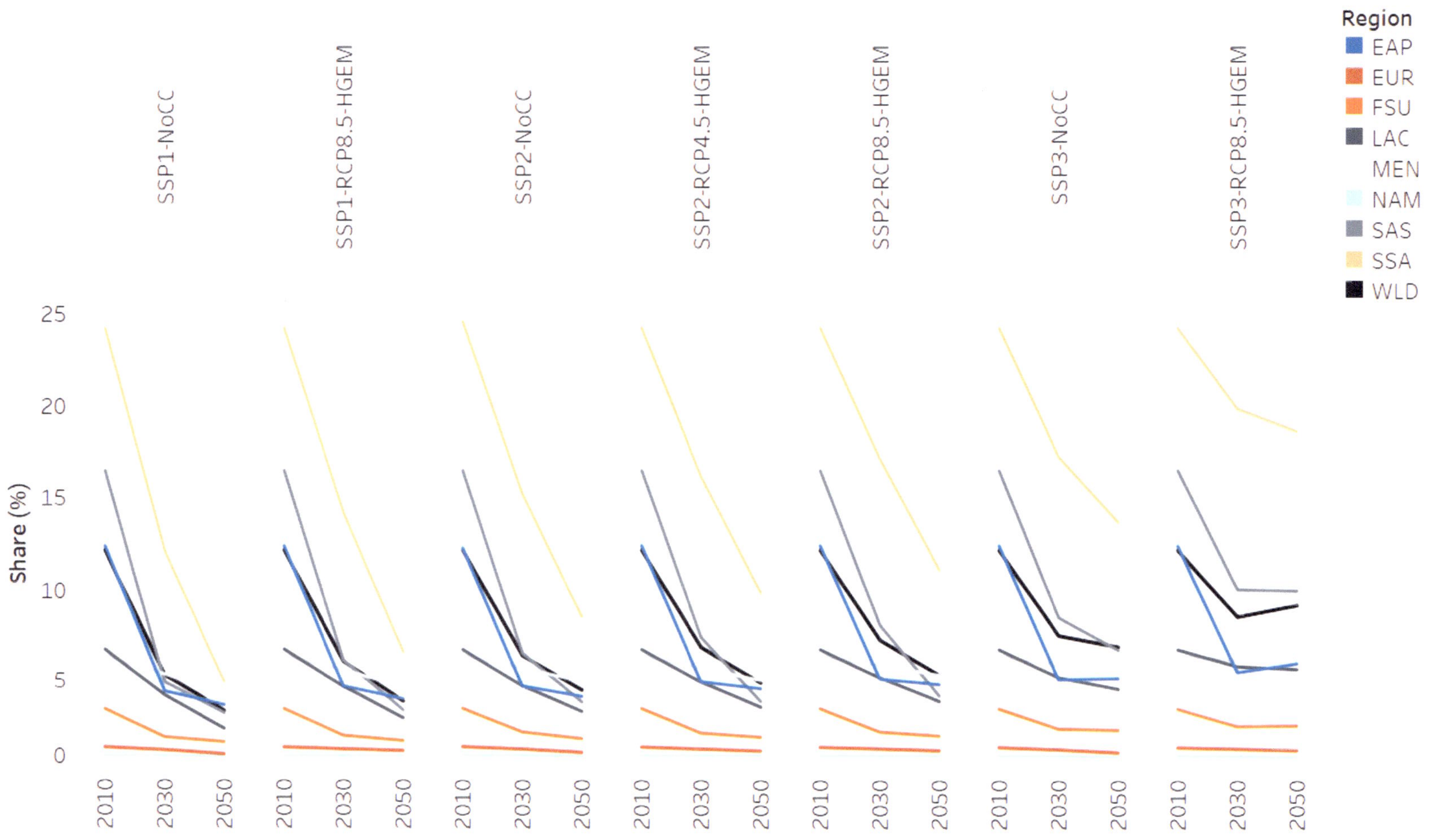

d) Across investment scenarios combined with SSP2-RCP8.5-HGEM in 2050

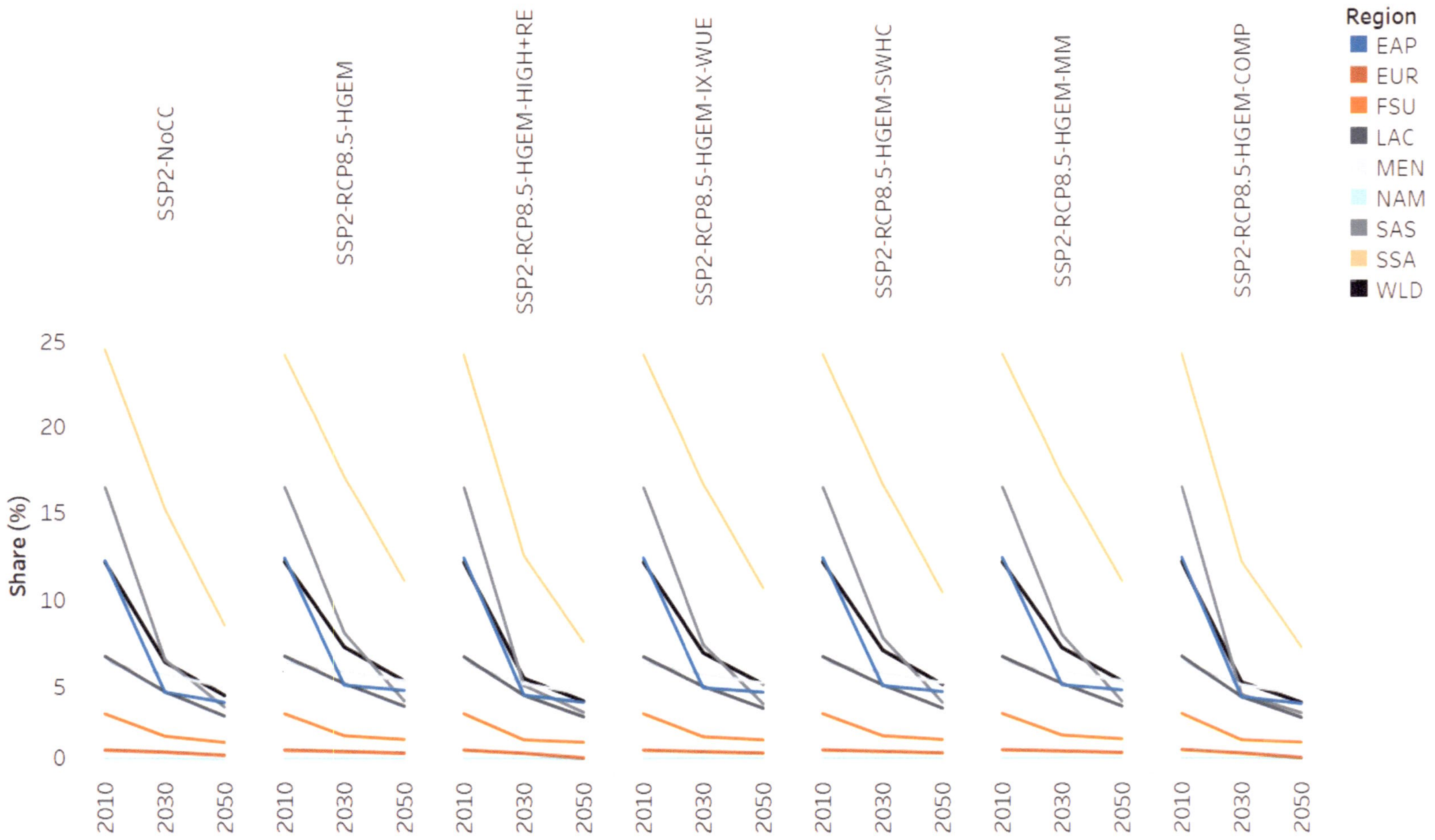

Regions are EAP-East Asia and Pacific; EUR-Europe; FSU-Former Soviet Union; LAC-Latin America and the Caribbean; MEN-Middle East and North Africa; NAM-North America; SAS-South Asia; SSA-Africa south of the Sahara; WLD-World.

Appendix C. Extended IMPACT model results for reference and investment scenarios

World commodity price tables

Table C – 1 World prices for animal products (2005 US$ per metric ton)

Scenario	Beef			Lamb			Pork			Poultry			Dairy			Eggs		
	2010	2030	2050	2010	2030	2050	2010	2030	2050	2010	2030	2050	2010	2030	2050	2010	2030	2050
REF-CC	3236	3958	3907	4044	5056	4360	2080	2771	2831	1569	2099	2094	495	568	570	1890	2164	2109
HIGH+RE		-21.5	-27.7		-24.5	-31.5		-5.9	-7.9		-19.0	-23.9		-10.4	-13.8		-1.2	-1.7
IX+WUE		-0.3	-0.2		0.1	0.1		-0.4	-0.4		-0.4	-0.4		-0.1	-0.1		-0.3	-0.3
SWHC		-0.2	-0.2		0.0	0.2		-0.2	-0.5		-0.4	-0.8		-0.1	-0.2		-0.3	-0.6
RMM		-5.6	-6.3		-4.7	-5.2		-5.4	-5.5		-4.5	-5.3		-2.5	-2.9		-5.9	-6.4
COMP		-26.1	-32.4		-27.9	-34.8		-11.0	-13.1		-22.8	-28.1		-12.9	-16.4		-7.1	-8.0

Note: Solution value for REF-CC – all other scenarios show percent change from REF-CC

Table C – 2 World prices for cereals (2005 US$ per metric ton)

Scenario	Barley			Maize			Millet			Rice			Sorghum			Wheat		
	2010	2030	2050	2010	2030	2050	2010	2030	2050	2010	2030	2050	2010	2030	2050	2010	2030	2050
REF-CC	172	187	198	149	215	301	282	364	413	359	455	567	147	183	194	221	261	317
HIGH+RE		-31.2	-38.2		-4.9	-10.6		-32.1	-37.5		-21.4	-26.6		-25.6	-31.6		-22.7	-29.3
IX+WUE		-1.7	-1.8		-1.5	-1.3		-0.7	-1.0		-7.4	-6.9		-0.9	-1.2		-3.6	-3.6
SWHC		-1.6	-3.1		-0.5	-2.5		-5.0	-8.3		-2.2	-3.7		-1.2	-2.5		-0.8	-3.5
RMM		-5.2	-6.4		-1.2	-2.3		-5.1	-5.6		-5.2	-6.0		-3.8	-4.3		-4.1	-5.8
COMP		-35.4	-42.7		-5.5	-12.9		-38.9	-46.0		-31.1	-36.2		-29.9	-36.6		-28.4	-34.6

Note: Solution value for REF-CC – all other scenarios show percent change from REF-CC

Table C – 3 World prices for major oilseeds and food oils (2005 US$ per metric ton)

Scenario	Groundnuts			Groundnut Oil			Soybeans			Soybean Oil			Palm Oil			Palm Kernel Oil		
	2010	2030	2050	2010	2030	2050	2010	2030	2050	2010	2030	2050	2010	2030	2050	2010	2030	2050
REF-CC	788	1030	1283	1183	1418	1540	333	449	517	615	743	764	522	657	687	574	692	799
HIGH+RE		-37.0	-43.1		-3.5	-5.6		-1.4	-2.7		0.9	0.9		-4.8	-5.5		-0.4	-0.9
IX+WUE		-2.1	-2.8		-0.7	-1.0		2.5	2.1		0.2	0.2		0.1	0.1		0.1	0.1
SWHC		-2.1	-3.5		-0.5	-0.8		-1.5	-2.8		-0.5	-0.8		-0.4	-0.4		-0.0	0.0
RMM		3.4	2.8		0.5	0.4		1.4	0.9		0.3	0.2		1.1	0.8		0.3	0.2
COMP		-37.4	-44.7		-4.2	-6.6		1.4	-1.3		1.0	0.8		-4.0	-4.9		0.0	-0.6

Note: Solution value for REF-CC – all other scenarios show percent change from REF-CC

Table C – 4 World prices for pulses and fruits & vegetables (2005 US$ per metric ton)

Scenario	Beans			Chickpeas			Cowpeas			Lentils			Pigeon peas			Fruits & Vegetables		
	2010	2030	2050	2010	2030	2050	2010	2030	2050	2010	2030	2050	2010	2030	2050	2010	2030	2050
REF-CC	1717	2118	2290	550	689	796	646	849	969	531	588	584	468	579	650	869	1086	1311
HIGH+RE		-26.8	-32.0		-29.4	-33.6		-25.6	-30.0		-25.4	-27.4		-31.2	-36.2		-5.3	-6.3
IX+WUE		-0.2	-0.4		-2.0	-2.3		-0.7	-1.0		-1.6	-1.6		-0.4	-0.6		-1.7	-1.9
SWHC		-1.9	-3.2		-1.3	-2.6		-3.8	-5.8		-0.8	-1.8		-1.5	-3.0		-1.3	-2.4
RMM		-5.2	-5.8		-5.5	-5.8		-7.1	-7.5		-4.0	-4.1		-6.0	-6.4		-5.4	-5.9
COMP		-31.6	-37.5		-35.0	-39.7		-33.9	-39.5		-30.1	-32.2		-36.0	-41.6		-12.6	-14.6

Note: Solution value for REF-CC – all other scenarios show percent change from REF-CC

Table C – 5 World prices for bananas, roots, and tubers (2005 US$ per metric ton)

Scenario	Banana			Plantains			Cassava			Potato			Sweet Potato			Yams		
	2010	2030	2050	2010	2030	2050	2010	2030	2050	2010	2030	2050	2010	2030	2050	2010	2030	2050
REF-CC	569	737	824	576	687	809	117	140	165	269	384	439	694	815	958	692	823	1035
HIGH+RE		-32.3	-37.2		-32.0	-37.2		-26.4	-31.4		-17.6	-20.9		-32.5	-37.5		-28.5	-33.8
IX+WUE		-2.8	-3.2		-0.8	-1.4		0.7	0.4		-1.9	-2.4		-1.1	-2.1		-0.4	-0.5
SWHC		-2.2	-3.5		-6.8	-10.8		-1.9	-3.7		-0.8	-1.6		-2.0	-3.8		-3.2	-5.2
RMM		-5.8	-6.3		-6.3	-6.7		-3.3	-3.9		-5.1	-5.7		-4.6	-5.4		-4.9	-5.4
COMP		-38.8	-44.3		-40.9	-48.3		-29.2	-35.5		-23.6	-27.9		-37.1	-43.7		-34.4	-41.1

Note: Solution value for REF-CC – all other scenarios show percent change from REF-CC

Crop yield tables

Table C – 6 Maize yields by year and region (metric tons per hectare)

Scenario	World			Developing			East Asia and Pacific			South Asia			Africa south of the Sahara			Middle East and North Africa			Latin America and Caribbean		
	2010	2030	2050	2010	2030	2050	2010	2030	2050	2010	2030	2050	2010	2030	2050	2010	2030	2050	2010	2030	2050
REF-CC	5.0	5.3	5.5	3.5	4.1	4.5	4.9	5.8	6.4	2.2	2.6	2.9	1.6	2.0	2.2	6.4	7.0	6.9	3.5	4.3	4.7
HIGH+RE		11.1	14.2		21.2	24.8		11.4	12.3		38.7	46.6		38.1	45.5		39.9	49.8		31.7	37.6
IX+WUE		1.6	1.2		2.9	2.0		4.3	1.8		2.3	1.9		0.8	1.5		10.3	12.4		1.0	1.6
SWHC		0.8	1.5		1.5	2.9		1.3	2.8		1.2	1.4		3.7	6.8		0.1	0.2		1.5	2.8
RMM		0.2	0.2		1.4	1.4		1.2	1.3		1.8	1.7		2.1	2.0		0.7	0.4		1.6	1.4
COMP		14.8	19.0		28.3	32.8		19.1	19.1		46.3	54.3		47.9	61.5		54.5	67.2		37.5	45.8

Note: Solution value for REF-CC – all other scenarios show percent change from REF-CC

Table C – 7 Barley yields by year and region (metric tons per hectare)

Scenario	World			Developing			East Asia and Pacific			South Asia			Africa south of the Sahara			Middle East and North Africa			Latin America and Caribbean		
	2010	2030	2050	2010	2030	2050	2010	2030	2050	2010	2030	2050	2010	2030	2050	2010	2030	2050	2010	2030	2050
REF-CC	2.7	3.5	4.2	2.1	2.9	3.5	4.3	5.7	5.9	1.8	2.3	2.5	1.6	2.2	2.6	1.8	2.3	2.4	2.9	4.0	4.4
HIGH+RE		11.8	13.9		26.6	31.6		29.3	35.6		27.7	35.0		24.3	29.3		32.8	39.7		26.8	32.0
IX+WUE		0.0	0.0		0.1	0.1		-0.2	-0.2		0.5	0.5		0.3	0.5		0.3	0.4		-0.6	-0.9
SWHC		-0.0	-0.1		0.1	0.1		-0.1	-0.2		-0.0	-0.1		1.1	2.3		0.3	0.7		0.4	0.7
RMM		-0.1	-0.3		0.7	0.6		0.8	0.6		1.0	0.8		0.8	0.6		0.5	0.4		0.9	0.8
COMP		12.3	14.3		27.9	33.1		30.4	36.9		30.0	37.2		27.6	34.4		34.7	42.2		27.5	33.1

Note: Solution value for REF-CC – all other scenarios show percent change from REF-CC

Table C – 8 Rice yields by year and region (metric tons per hectare)

Scenario	World			Developing			East Asia and Pacific			South Asia			Africa south of the Sahara			Middle East and North Africa			Latin America and Caribbean		
	2010	2030	2050	2010	2030	2050	2010	2030	2050	2010	2030	2050	2010	2030	2050	2010	2030	2050	2010	2030	2050
REF-CC	2.8	3.2	3.4	2.8	3.2	3.3	3.3	3.6	3.9	2.3	2.7	2.6	1.3	2.1	2.6	4.7	5.2	5.9	3.0	3.6	3.7
HIGH+RE		9.3	11.4		9.9	12.3		8.4	9.8		9.3	10.9		32.9	39.6		9.6	11.1		8.0	8.7
IX+WUE		1.5	1.0		1.7	1.2		1.6	0.9		1.4	1.1		2.6	3.5		5.1	4.5		3.4	2.9
SWHC		0.5	0.9		0.6	1.0		0.7	1.2		0.2	0.3		2.0	3.2		-0.1	-0.1		0.2	0.1
RMM		0.7	0.6		0.8	0.7		0.6	0.6		0.8	0.7		0.9	0.8		0.7	0.8		1.2	1.1
COMP		12.0	14.0		12.9	15.2		11.1	11.9		11.9	13.1		40.5	50.2		14.1	15.4		13.0	13.4

Note: Solution value for REF-CC – all other scenarios show percent change from REF-CC

Table C – 9 Wheat yields by year and region (metric tons per hectare)

Scenario	World			Developing			East Asia and Pacific			South Asia			Africa south of the Sahara			Middle East and North Africa			Latin America and Caribbean		
	2010	2030	2050	2010	2030	2050	2010	2030	2050	2010	2030	2050	2010	2030	2050	2010	2030	2050	2010	2030	2050
REF-CC	3.0	3.6	4.1	2.7	3.3	3.8	4.6	5.4	5.4	2.7	3.6	4.5	2.2	2.7	3.1	2.4	3.1	3.6	2.7	3.3	3.7
HIGH+RE		14.5	17.0		23.5	27.2		33.2	40.5		23.5	27.2		32.6	39.2		30.1	36.5		35.6	43.2
IX+WUE		1.5	1.6		2.5	2.7		0.4	0.3		6.4	5.9		0.6	0.2		1.1	1.2		1.0	0.5
SWHC		0.2	0.0		0.3	0.3		0.0	-0.3		0.6	0.8		1.6	3.1		0.2	0.3		0.7	1.1
RMM		0.6	0.3		1.3	1.0		1.3	1.0		2.0	1.5		1.4	1.1		0.9	0.6		1.1	0.8
COMP		17.9	21.0		28.6	33.2		35.7	42.9		34.6	38.4		37.6	46.3		33.3	40.2		39.6	47.3

Note: Solution value for REF-CC – all other scenarios show percent change from REF-CC

Table C – 10 Millet yields by year and region (metric tons per hectare)

Scenario	World			Developing			East Asia and Pacific			South Asia			Africa south of the Sahara			Middle East and North Africa			Latin America and Caribbean		
	2010	2030	2050	2010	2030	2050	2010	2030	2050	2010	2030	2050	2010	2030	2050	2010	2030	2050	2010	2030	2050
REF-CC	0.9	1.1	1.3	0.9	1.1	1.3	1.9	2.2	2.6	1.0	1.1	1.2	0.9	1.1	1.3	0.9	1.6	2.1	1.7	2.0	2.4
HIGH+RE		21.0	25.5		21.3	25.9		-0.3	-0.4		24.3	29.4		23.6	28.6		13.0	12.9		2.2	2.5
IX+WUE		0.1	0.1		0.1	0.1		0.0	0.0		0.0	0.0		0.0	0.0		8.5	4.5		-0.1	-0.1
SWHC		1.7	3.8		1.8	3.9		1.0	2.1		0.9	1.8		2.4	4.9		-0.2	-0.0		1.8	3.7
RMM		0.8	0.7		0.8	0.7		0.7	0.7		1.0	0.9		0.8	0.7		0.1	0.1		0.5	0.5
COMP		24.3	31.5		24.7	32.0		1.5	2.6		26.7	33.0		27.6	36.0		19.7	14.6		4.5	6.8

Note: Solution value for REF-CC – all other scenarios show percent change from REF-CC

Table C – 11 Sorghum yields by year and region (metric tons per hectare)

Scenario	World			Developing			East Asia and Pacific			South Asia			Africa south of the Sahara			Middle East and North Africa			Latin America and Caribbean		
	2010	2030	2050	2010	2030	2050	2010	2030	2050	2010	2030	2050	2010	2030	2050	2010	2030	2050	2010	2030	2050
REF-CC	1.4	1.7	2.0	1.2	1.5	1.8	4.7	5.6	6.3	0.8	1.0	1.0	1.0	1.3	1.5	1.5	1.7	1.8	3.3	3.9	4.5
HIGH+RE		15.8	20.7		20.9	26.7		8.0	9.3		26.2	31.7		34.2	41.3		24.7	29.8		1.2	1.3
IX+WUE		0.3	0.3		0.3	0.3		-0.3	-0.3		0.0	0.0		0.8	0.8		1.5	3.6		0.3	0.0
SWHC		0.5	1.1		0.7	1.4		0.2	0.3		0.2	0.4		0.9	1.8		-0.0	0.0		0.4	0.7
RMM		0.6	0.6		1.2	1.1		1.0	0.9		1.2	1.1		1.0	0.9		0.1	0.5		1.1	1.0
COMP		17.7	23.4		23.6	30.4		9.2	10.8		27.9	33.7		37.8	46.2		27.1	35.1		3.0	3.1

Note: Solution value for REF-CC – all other scenarios show percent change from REF-CC

Table C – 12 Banana yields by year and region (metric tons per hectare)

Scenario	World			Developing			East Asia and Pacific			South Asia			Africa south of the Sahara			Middle East and North Africa			Latin America and Caribbean		
	2010	2030	2050	2010	2030	2050	2010	2030	2050	2010	2030	2050	2010	2030	2050	2010	2030	2050	2010	2030	2050
REF-CC	17.7	21.2	22.7	17.6	21.1	22.7	17.6	20.1	21.4	29.6	40.1	46.5	7.7	9.4	10.2	34.8	45.8	50.7	21.7	24.3	24.3
HIGH+RE		20.2	23.9		20.4	24.1		22.7	27.7		23.0	27.0		20.0	21.9		1.8	2.1		18.9	22.1
IX+WUE		0.7	0.8		0.7	0.8		-0.1	-0.2		0.2	0.1		1.5	2.1		-0.2	-0.4		2.0	2.6
SWHC		0.6	0.9		0.6	0.9		0.4	0.9		0.2	0.1		3.0	5.0		-0.3	-0.3		0.4	0.6
RMM		1.2	1.1		1.2	1.2		1.1	1.0		1.7	1.6		1.0	0.8		1.0	1.0		0.8	0.6
COMP		23.0	27.2		23.3	27.4		24.6	30.1		25.7	29.7		26.4	32.0		2.5	2.6		21.6	25.8

Note: Solution value for REF-CC – all other scenarios show percent change from REF-CC

Table C – 13 Plantain yields by year and region (metric tons per hectare)

Scenario	World			Developing			East Asia and Pacific			South Asia			Africa south of the Sahara			Middle East and North Africa			Latin America and Caribbean		
	2010	2030	2050	2010	2030	2050	2010	2030	2050	2010	2030	2050	2010	2030	2050	2010	2030	2050	2010	2030	2050
REF-CC	6.7	9.1	10.6	6.7	9.1	10.6	11.7	16.1	19.4	12.3	16.4	19.3	6.0	8.2	9.7	0.0	0.0	0.0	9.4	12.5	14.5
HIGH+RE		17.8	21.3		17.8	21.3		5.8	6.8		18.7	22.6		19.0	22.8					15.6	18.5
IX+WUE		0.2	0.4		0.2	0.4		0.0	-0.1		0.1	0.0		-0.1	-0.1					2.2	2.9
SWHC		2.6	5.6		2.6	5.6		-0.1	0.2		-0.1	0.3		3.8	7.9					-0.5	-0.9
RMM		1.2	1.0		1.2	1.0		0.9	0.9		1.3	1.3		1.1	1.0					1.2	1.1
COMP		22.6	30.0		22.6	30.0		6.6	7.9		20.3	24.5		24.8	33.6					18.8	21.9

Note: Solution value for REF-CC – all other scenarios show percent change from REF-CC

Table C – 14 Cassava yields by year and region (metric tons per hectare)

Scenario	World			Developing			East Asia and Pacific			South Asia			Africa south of the Sahara			Middle East and North Africa			Latin America and Caribbean		
	2010	2030	2050	2010	2030	2050	2010	2030	2050	2010	2030	2050	2010	2030	2050	2010	2030	2050	2010	2030	2050
REF-CC	12.2	14.6	15.9	12.2	14.6	15.9	18.4	21.1	21.8	25.3	28.0	26.8	10.1	12.4	14.1	0.0	0.0	0.0	13.5	17.2	18.4
HIGH+RE		21.0	25.1		21.0	25.1		21.9	26.5		21.0	25.4		20.2	24.3					21.3	25.8
IX+WUE		-0.1	-0.1		-0.1	-0.1		0.1	0.2		0.3	0.2		0.1	0.1					-0.2	-0.2
SWHC		0.8	1.7		0.8	1.7		0.4	0.8		0.1	0.1		1.2	2.2					0.4	1.0
RMM		1.0	0.9		1.0	0.9		0.8	0.8		1.1	1.0		1.1	1.0					1.0	1.0
COMP		23.2	28.5		23.2	28.5		23.7	28.9		22.9	27.2		23.1	28.6					23.0	28.1

Note: Solution value for REF-CC – all other scenarios show percent change from REF-CC

Table C – 15 Potato yields by year and region (metric tons per hectare)

Scenario	World			Developing			East Asia and Pacific			South Asia			Africa south of the Sahara			Middle East and North Africa			Latin America and Caribbean		
	2010	2030	2050	2010	2030	2050	2010	2030	2050	2010	2030	2050	2010	2030	2050	2010	2030	2050	2010	2030	2050
REF-CC	17.9	18.6	19.5	15.4	16.2	17.1	14.9	16.2	15.3	17.9	20.8	24.7	8.7	8.9	8.5	24.8	29.1	32.6	17.1	19.9	21.8
HIGH+RE		8.1	9.8		12.4	14.6		-0.7	-1.0		5.5	6.4		15.8	19.1		34.6	41.3		33.7	41.1
IX+WUE		0.2	0.2		0.4	0.6		-0.5	-0.7		-0.2	-0.3		3.1	4.4		1.8	1.7		3.1	3.2
SWHC		0.2	0.2		0.3	0.4		0.8	1.4		-0.1	-0.3		1.6	1.6		0.0	-0.1		0.2	0.4
RMM		1.0	1.0		2.0	1.9		2.0	1.8		2.5	2.3		2.4	2.2		1.4	1.3		2.0	1.8
COMP		9.9	11.9		15.5	18.2		1.6	1.9		7.9	8.3		23.9	28.5		38.8	46.2		40.3	48.4

Note: Solution value for REF-CC – all other scenarios show percent change from REF-CC

Table C – 16 Sweet Potato yields by year and region (metric tons per hectare)

Scenario	World			Developing			East Asia and Pacific			South Asia			Africa south of the Sahara			Middle East and North Africa			Latin America and Caribbean		
	2010	2030	2050	2010	2030	2050	2010	2030	2050	2010	2030	2050	2010	2030	2050	2010	2030	2050	2010	2030	2050
REF-CC	14.3	16.3	16.7	14.2	16.1	16.5	21.1	24.5	25.4	9.1	11.2	11.9	4.7	6.9	8.9	16.7	21.8	23.0	9.1	12.7	15.8
HIGH+RE		18.9	21.6		19.4	22.3		19.6	23.3		14.4	17.3		17.5	18.3		-3.0	-3.6		20.4	24.6
IX+WUE		-0.4	-0.5		-0.4	-0.5		-0.2	-0.4		0.0	-0.3		1.1	1.3		-0.1	-0.3		0.3	0.2
SWHC		0.7	1.5		0.8	1.5		0.6	1.3		-0.2	-0.3		1.3	2.4		-0.4	-0.7		0.6	1.2
RMM		1.6	1.4		1.7	1.5		1.7	1.5		1.9	1.8		1.8	1.5		1.5	1.3		1.6	1.5
COMP		21.3	24.7		21.9	25.5		22.3	26.6		16.6	18.9		21.7	23.9		-1.9	-3.1		23.8	28.3

Note: Solution value for REF-CC – all other scenarios show percent change from REF-CC

Table C – 17 Yam yields by year and region (metric tons per hectare)

Scenario	World			Developing			East Asia and Pacific			South Asia			Africa south of the Sahara			Middle East and North Africa			Latin America and Caribbean		
	2010	2030	2050	2010	2030	2050	2010	2030	2050	2010	2030	2050	2010	2030	2050	2010	2030	2050	2010	2030	2050
REF-CC	12.0	18.3	24.7	12.0	18.3	24.7	12.4	18.2	24.8	0.0	0.0	0.0	12.1	18.4	24.9	7.4	11.5	15.7	9.0	13.6	18.2
HIGH+RE		20.2	24.3		20.3	24.4		-2.9	-3.8					20.6	24.6		-8.4	-10.2		16.4	19.5
IX+WUE		0.0	0.0		0.0	0.0		-0.1	-0.1					0.0	0.0		-0.1	-0.1		0.1	0.1
SWHC		1.5	3.0		1.5	3.0		-0.6	-1.0					1.5	3.1		1.4	2.7		0.2	0.8
RMM		1.5	1.3		1.5	1.4		1.8	1.6					1.5	1.3		1.6	1.5		1.7	1.6
COMP		23.9	29.7		23.9	29.7		-1.9	-3.5					24.2	30.1		-5.6	-6.6		18.8	22.4

Note: Solution value for REF-CC – all other scenarios show percent change from REF-CC

Table C – 18 Bean yields by year and region (metric tons per hectare)

Scenario	World			Developing			East Asia and Pacific			South Asia			Africa south of the Sahara			Middle East and North Africa			Latin America and Caribbean		
	2010	2030	2050	2010	2030	2050	2010	2030	2050	2010	2030	2050	2010	2030	2050	2010	2030	2050	2010	2030	2050
REF-CC	0.8	1.0	1.2	0.7	1.0	1.1	1.2	1.5	1.8	0.4	0.5	0.5	0.6	0.8	0.9	1.9	2.1	2.1	0.9	1.1	1.3
HIGH+RE		19.1	23.1		23.4	28.8		16.6	19.2		16.0	19.1		19.2	22.8		16.8	19.5		31.2	37.8
IX+WUE		0.9	0.9		0.8	0.8		0.4	0.4		0.4	0.4		0.7	0.8		1.6	1.5		3.0	2.9
SWHC		0.6	1.2		0.7	1.5		0.4	0.7		0.1	0.3		2.1	4.0		-0.1	-0.2		0.5	1.2
RMM		1.0	0.9		1.7	1.7		1.5	1.5		1.7	1.5		2.1	1.9		1.4	1.1		2.1	2.1
COMP		22.3	27.4		27.3	34.1		19.4	22.7		18.6	22.1		25.2	31.4		20.2	22.7		38.8	46.7

Note: Solution value for REF-CC – all other scenarios show percent change from REF-CC

Table C – 19 Chickpea yields by year and region (metric tons per hectare)

Scenario	World			Developing			East Asia and Pacific			South Asia			Africa south of the Sahara			Middle East and North Africa			Latin America and Caribbean		
	2010	2030	2050	2010	2030	2050	2010	2030	2050	2010	2030	2050	2010	2030	2050	2010	2030	2050	2010	2030	2050
REF-CC	0.9	1.2	1.5	0.9	1.2	1.4	1.4	1.6	1.6	0.8	1.2	1.5	0.8	1.1	1.4	0.9	1.1	1.1	1.4	2.1	2.7
HIGH+RE		19.1	22.2		20.9	25.2		22.7	27.7		19.4	23.7		28.3	34.4		31.7	38.8		-3.9	-4.6
IX+WUE		0.3	0.2		0.3	0.2		0.0	0.0		-0.2	-0.2		-0.5	-0.3		3.4	4.2		3.8	3.7
SWHC		0.3	0.6		0.3	0.7		0.2	0.2		0.3	0.6		1.4	2.7		0.4	0.5		0.4	0.8
RMM		1.2	1.0		1.4	1.3		1.1	1.0		1.7	1.6		1.8	1.6		0.9	0.7		-1.4	-1.4
COMP		21.6	25.1		23.7	28.6		24.7	30.4		21.7	26.5		32.0	40.2		37.2	45.8		-1.4	-1.6

Note: Solution value for REF-CC – all other scenarios show percent change from REF-CC

Table C – 20 Cowpea yields by year and region (metric tons per hectare)

Scenario	World			Developing			East Asia and Pacific			South Asia			Africa south of the Sahara			Middle East and North Africa			Latin America and Caribbean		
	2010	2030	2050	2010	2030	2050	2010	2030	2050	2010	2030	2050	2010	2030	2050	2010	2030	2050	2010	2030	2050
REF-CC	0.5	0.8	1.0	0.5	0.7	1.0	1.1	1.3	1.3	1.0	1.4	1.7	0.5	0.7	1.0	0.7	0.9	1.2	0.9	1.4	1.9
HIGH+RE		19.3	23.9		19.6	24.2		-5.3	-6.3		-6.9	-8.2		20.7	25.3		-10.2	-11.7		-7.6	-9.4
IX+WUE		-0.1	0.0		-0.1	0.0		0.2	0.2		-0.2	-0.2		-0.1	-0.1		0.1	0.2		3.2	3.3
SWHC		1.5	3.3		1.6	3.3		-0.0	0.2		1.2	2.4		1.6	3.4		-0.0	0.1		-0.3	-0.3
RMM		1.4	1.3		1.4	1.3		0.7	0.7		1.7	1.6		1.5	1.4		1.1	1.1		1.3	1.2
COMP		22.8	29.5		23.2	30.0		-4.4	-5.4		-4.3	-4.7		24.4	31.2		-9.0	-10.6		-4.0	-5.8

Note: Solution value for REF-CC – all other scenarios show percent change from REF-CC

Table C – 21 Lentil yields by year and region (metric tons per hectare)

Scenario	World			Developing			East Asia and Pacific			South Asia			Africa south of the Sahara			Middle East and North Africa			Latin America and Caribbean		
	2010	2030	2050	2010	2030	2050	2010	2030	2050	2010	2030	2050	2010	2030	2050	2010	2030	2050	2010	2030	2050
REF-CC	1.0	1.4	1.7	0.9	1.1	1.2	1.8	2.5	3.1	0.7	0.9	1.1	0.7	1.0	1.2	1.1	1.3	1.4	0.8	1.0	1.3
HIGH+RE		13.1	12.1		29.9	37.0		22.0	27.6		29.6	37.2		20.8	26.2		32.8	40.9		17.5	22.4
IX+WUE		0.1	0.0		0.5	0.6		-0.3	-0.3		-0.3	-0.3		-0.4	-0.4		1.5	2.0		1.2	1.3
SWHC		0.1	0.1		0.4	0.7		0.5	0.7		0.3	0.6		1.7	3.8		0.2	0.2		0.7	1.4
RMM		0.2	-0.3		2.0	1.9		2.2	2.2		2.0	2.0		2.6	2.6		1.8	1.7		0.4	0.8
COMP		14.5	13.2		33.6	41.6		25.4	32.1		32.2	40.5		25.5	34.0		37.2	46.5		19.0	25.5

Note: Solution value for REF-CC – all other scenarios show percent change from REF-CC

Table C – 22 Pigeonpea yields by year and region (metric tons per hectare)

Scenario	World			Developing			East Asia and Pacific			South Asia			Africa south of the Sahara			Middle East and North Africa			Latin America and Caribbean		
	2010	2030	2050	2010	2030	2050	2010	2030	2050	2010	2030	2050	2010	2030	2050	2010	2030	2050	2010	2030	2050
REF-CC	0.8	1.0	1.2	0.8	1.0	1.2	1.1	1.3	1.4	0.7	1.0	1.2	0.7	0.9	1.1	0.0	0.0	0.0	0.9	1.3	1.6
HIGH+RE		21.3	26.5		21.3	26.5		-2.3	-2.8		26.9	32.8		16.2	19.2					-8.5	-10.3
IX+WUE		0.0	0.0		0.0	0.0		0.2	0.2		0.1	0.1		0.0	0.0					0.5	0.5
SWHC		0.4	0.9		0.4	0.9		0.4	0.7		0.2	0.5		1.9	3.4					0.6	1.2
RMM		1.4	1.3		1.4	1.3		0.9	0.9		1.5	1.4		1.7	1.5					1.4	1.3
COMP		23.9	29.8		23.9	29.8		-0.7	-0.8		29.4	35.9		20.6	25.8					-6.1	-7.4

Note: Solution value for REF-CC – all other scenarios show percent change from REF-CC

Table C – 23 Groundnut yields by year and region (metric tons per hectare)

Scenario	World			Developing			East Asia and Pacific			South Asia			Africa south of the Sahara			Middle East and North Africa			Latin America and Caribbean		
	2010	2030	2050	2010	2030	2050	2010	2030	2050	2010	2030	2050	2010	2030	2050	2010	2030	2050	2010	2030	2050
REF-CC	1.5	1.6	1.6	1.5	1.6	1.5	2.6	2.9	3.1	1.0	1.0	0.9	1.0	1.1	1.1	3.1	3.4	3.2	2.2	2.3	2.2
HIGH+RE		12.2	15.0		14.1	17.2		12.5	14.2		10.0	10.9		21.6	26.2		-2.3	-3.8		-2.0	-2.2
IX+WUE		0.7	0.7		0.8	0.8		0.2	0.1		1.6	1.1		1.9	2.4		1.0	1.6		0.4	0.3
SWHC		0.6	1.2		0.7	1.3		0.3	0.5		0.6	0.8		1.3	2.6		-0.1	-0.3		0.7	1.5
RMM		0.4	0.4		0.2	0.2		0.1	0.0		0.3	0.3		0.4	0.3		0.1	0.0		0.3	0.2
COMP		14.2	17.7		16.1	19.9		13.1	14.7		12.6	12.8		26.1	33.0		-1.1	-1.8		-0.6	-0.2

Note: Solution value for REF-CC – all other scenarios show percent change from REF-CC

Table C – 24 Soybean yields by year and region (metric tons per hectare)

Scenario	World			Developing			East Asia and Pacific			South Asia			Africa south of the Sahara			Middle East and North Africa			Latin America and Caribbean		
	2010	2030	2050	2010	2030	2050	2010	2030	2050	2010	2030	2050	2010	2030	2050	2010	2030	2050	2010	2030	2050
REF-CC	2.4	2.6	2.5	2.1	2.3	2.3	1.7	1.9	1.9	1.0	1.1	0.9	1.1	1.3	1.3	2.9	2.3	1.8	2.5	2.7	2.7
HIGH+RE		0.1	0.1		-0.1	-0.2		-0.3	-0.5		0.3	0.3		21.6	23.6		-2.7	-0.4		0.0	-0.1
IX+WUE		0.6	0.6		0.4	0.5		1.5	1.1		1.8	2.2		0.8	1.2		21.8	28.8		0.7	0.9
SWHC		0.6	1.4		1.1	2.3		1.0	2.5		0.6	1.0		3.5	6.0		2.3	0.2		1.1	2.4
RMM		0.4	0.4		0.3	0.2		0.3	0.2		0.3	0.2		0.2	0.1		-1.2	0.2		0.2	0.1
COMP		1.6	2.5		1.7	3.0		2.4	3.3		3.1	4.1		27.2	32.5		21.8	28.9		2.0	3.4

Note: Solution value for REF-CC – all other scenarios show percent change from REF-CC

Table C – 25 Palm Fruit yields by year and region (metric tons per hectare)

Scenario	World			Developing			East Asia and Pacific			South Asia			Africa south of the Sahara			Middle East and North Africa			Latin America and Caribbean		
	2010	2030	2050	2010	2030	2050	2010	2030	2050	2010	2030	2050	2010	2030	2050	2010	2030	2050	2010	2030	2050
REF-CC	15.8	18.7	19.9	15.8	18.7	19.9	19.8	22.1	22.5	0.0	0.0	0.0	7.6	9.9	10.9	0.0	0.0	0.0	16.2	19.0	19.1
HIGH+RE		10.9	12.9		10.9	12.9		10.7	12.8					10.2	12.3					10.5	12.8
IX+WUE		0.6	0.7		0.6	0.7		0.9	1.1					0.0	0.0					1.1	1.4
SWHC		0.3	0.6		0.3	0.6		-0.0	0.0					2.8	5.3					-0.1	0.0
RMM		0.3	0.2		0.3	0.2		0.2	0.1					0.3	0.2					0.3	0.2
COMP		12.2	14.7		12.2	14.7		11.8	14.2					13.7	18.5					12.0	14.6

Note: Solution value for REF-CC – all other scenarios show percent change from REF-CC

Commodity production tables

Table C – 26 Beef production by year and region (million metric tons)

Scenario	World			Developing			East Asia and Pacific			South Asia			Africa south of the Sahara			Middle East and North Africa			Latin America and Caribbean		
	2010	2030	2050	2010	2030	2050	2010	2030	2050	2010	2030	2050	2010	2030	2050	2010	2030	2050	2010	2030	2050
REF-CC	69.6	98.2	121.0	43.4	67.9	89.6	8.2	14.1	17.3	4.7	7.6	10.8	5.2	9.5	15.7	2.3	4.3	7.5	18.7	27.2	33.0
HIGH+RE		5.6	7.7		12.1	14.4		4.4	4.3		20.3	23.8		27.5	32.2		5.3	5.3		12.1	13.0
IX+WUE		0.2	0.2		0.2	0.2		0.2	0.2		0.3	0.3		0.0	0.1		0.1	0.2		0.1	0.2
SWHC		0.1	0.3		0.1	0.3		0.1	0.3		0.1	0.4		0.0	0.1		0.1	0.2		0.1	0.3
RMM		1.4	1.5		3.0	2.8		2.5	2.4		2.2	2.3		2.2	1.9		2.1	2.0		4.1	3.9
COMP		7.4	9.6		15.5	17.6		7.1	6.8		23.1	26.7		30.3	34.7		7.5	7.5		16.7	17.4

Note: Solution value for REF-CC – all other scenarios show percent change from REF-CC

Table C – 27 Lamb production by year and region (million metric tons)

Scenario	World			Developing			East Asia and Pacific			South Asia			Africa south of the Sahara			Middle East and North Africa			Latin America and Caribbean		
	2010	2030	2050	2010	2030	2050	2010	2030	2050	2010	2030	2050	2010	2030	2050	2010	2030	2050	2010	2030	2050
REF-CC	14.6	23.5	32.7	11.6	19.3	27.3	4.1	5.2	4.5	2.0	3.8	5.9	2.3	4.7	8.4	1.9	3.4	5.3	0.5	0.8	1.2
HIGH+RE		9.4	12.7		12.9	16.9		3.9	3.6		14.1	15.5		30.4	35.8		5.4	5.6		17.6	20.3
IX+WUE		0.1	0.1		0.1	0.1		0.1	0.1		0.1	0.1		0.0	0.1		0.2	0.2		0.1	0.1
SWHC		0.1	0.2		0.0	0.2		0.0	0.2		0.0	0.1		0.0	0.1		0.1	0.3		0.1	0.2
RMM		1.6	1.6		2.1	2.0		2.1	2.0		2.7	2.5		1.7	1.6		2.2	2.2		2.1	2.0
COMP		11.3	14.7		15.4	19.4		6.2	5.8		17.3	18.5		32.7	38.1		7.8	8.0		20.1	22.8

Note: Solution value for REF-CC – all other scenarios show percent change from REF-CC

Table C – 28 Pork production by year and region (million metric tons)

Scenario	World			Developing			East Asia and Pacific			South Asia			Africa south of the Sahara			Middle East and North Africa			Latin America and Caribbean		
	2010	2030	2050	2010	2030	2050	2010	2030	2050	2010	2030	2050	2010	2030	2050	2010	2030	2050	2010	2030	2050
REF-CC	104.8	125.1	130.3	66.2	77.4	76.3	54.3	59.7	52.5	0.6	1.3	2.4	1.1	1.9	3.0	0.0	0.0	0.0	6.7	10.3	13.8
HIGH+RE		-0.3	-0.3		-0.6	-0.9		-0.6	-0.8		-3.1	-4.1		-0.2	-0.4		0.5	0.6		-2.1	-2.6
IX+WUE		0.1	0.1		0.0	0.0		0.0	0.0		0.0	0.0		0.1	0.1		0.0	0.0		0.0	0.0
SWHC		0.1	0.2		-0.0	0.1		-0.0	0.1		0.0	0.0		0.1	0.1		0.3	0.7		-0.0	0.0
RMM		1.2	1.1		3.3	3.3		3.4	3.5		1.9	1.9		1.6	1.6		1.3	1.3		3.1	3.1
COMP		1.0	0.9		2.6	2.3		2.7	2.6		-1.2	-2.1		1.4	1.3		1.7	1.9		0.9	0.5

Note: Solution value for REF-CC – all other scenarios show percent change from REF-CC

Table C – 29 Poultry production by year and region (million metric tons)

Scenario	World			Developing			East Asia and Pacific			South Asia			Africa south of the Sahara			Middle East and North Africa			Latin America and Caribbean		
	2010	2030	2050	2010	2030	2050	2010	2030	2050	2010	2030	2050	2010	2030	2050	2010	2030	2050	2010	2030	2050
REF-CC	85.1	131.9	168.6	52.6	87.5	114.2	21.8	37.1	43.7	2.7	6.4	11.4	2.2	4.4	7.2	5.0	8.6	12.9	18.2	27.7	35.5
HIGH+RE		4.5	5.8		10.1	12.3		4.6	4.8		14.1	16.2		33.0	40.0		5.9	6.9		15.2	17.7
IX+WUE		0.3	0.3		0.2	0.3		0.2	0.2		0.1	0.2		0.3	0.2		0.3	0.4		0.2	0.3
SWHC		0.1	0.3		0.1	0.3		0.1	0.3		0.1	0.2		0.1	0.2		0.2	0.5		0.1	0.3
RMM		1.1	1.1		2.4	2.4		2.0	1.9		3.4	3.2		3.1	2.9		2.8	2.8		2.6	2.4
COMP		6.0	7.4		12.9	15.2		6.7	7.0		18.1	20.1		37.2	44.1		9.0	10.1		18.4	20.8

Note: Solution value for REF-CC – all other scenarios show percent change from REF-CC

Table C – 30 Dairy production by year and region (million metric tons)

Scenario	World			Developing			East Asia and Pacific			South Asia			Africa south of the Sahara			Middle East and North Africa			Latin America and Caribbean		
	2010	2030	2050	2010	2030	2050	2010	2030	2050	2010	2030	2050	2010	2030	2050	2010	2030	2050	2010	2030	2050
REF-CC	709	967	1124	408.1	613.9	764.7	42.2	83.6	111.3	151.2	254.2	338.6	27.5	42.8	60.0	35.7	57.7	78.0	75.3	99.7	111.9
HIGH+RE		4.4	6.1		12.6	15.0		4.2	4.0		17.5	20.2		29.7	35.1		3.1	2.7		15.6	17.6
IX+WUE		0.1	0.1		0.2	0.2		0.1	0.1		0.2	0.2		0.1	0.1		0.2	0.2		0.2	0.2
SWHC		0.0	0.2		0.1	0.2		-0.0	0.0		0.2	0.3		0.0	0.1		0.1	0.3		0.0	0.2
RMM		1.1	1.2		3.1	3.0		2.7	2.5		3.1	3.0		3.2	3.0		3.8	3.7		3.5	3.3
COMP		5.8	7.6		16.0	18.3		6.9	6.5		21.1	23.7		33.7	39.0		7.0	6.4		19.6	21.5

Note: Solution value for REF-CC – all other scenarios show percent change from REF-CC

Table C – 31 Egg production by year and region (million metric tons)

Scenario	World			Developing			East Asia and Pacific			South Asia			Africa south of the Sahara			Middle East and North Africa			Latin America and Caribbean		
	2010	2030	2050	2010	2030	2050	2010	2030	2050	2010	2030	2050	2010	2030	2050	2010	2030	2050	2010	2030	2050
REF-CC	65.8	84.5	94.2	49.3	66.0	74.4	30.3	40.3	42.1	4.0	7.1	10.6	1.6	2.1	2.6	2.9	4.3	5.7	6.5	8.1	9.4
HIGH+RE		0.4	0.8		0.4	0.7		0.2	0.6		-1.3	-1.7		1.3	1.9		3.3	4.7		-0.1	0.0
IX+WUE		0.2	0.2		0.2	0.2		0.2	0.2		0.0	0.1		0.1	0.1		0.3	0.4		0.0	0.1
SWHC		0.1	0.2		0.1	0.2		0.1	0.2		0.0	0.0		0.0	0.2		0.3	0.6		0.0	0.1
RMM		1.5	1.5		2.5	2.5		2.9	2.9		1.8	1.7		1.8	1.8		2.5	2.6		2.0	2.0
COMP		2.1	2.5		3.1	3.4		3.3	3.6		0.6	0.1		3.1	3.8		6.2	7.8		2.0	2.0

Note: Solution value for REF-CC – all other scenarios show percent change from REF-CC

Table C – 32 Barley production by year and region (million metric tons)

Scenario	World			Developing			East Asia and Pacific			South Asia			Africa south of the Sahara			Middle East and North Africa			Latin America and Caribbean		
	2010	2030	2050	2010	2030	2050	2010	2030	2050	2010	2030	2050	2010	2030	2050	2010	2030	2050	2010	2030	2050
REF-CC	153.8	200.5	240.7	71.4	106.3	127.8	3.9	5.4	5.3	1.8	2.4	2.5	2.2	3.4	4.4	20.8	28.2	29.8	3.5	4.9	5.5
HIGH+RE		3.9	3.2		23.2	26.2		24.4	28.0		20.8	26.2		17.5	20.1		29.5	34.3		15.0	16.0
IX+WUE		0.0	-0.1		-0.1	-0.1		-0.5	-0.5		2.1	2.0		0.7	1.0		-0.5	-0.5		1.2	0.4
SWHC		-0.2	-0.6		-0.0	-0.4		-0.2	-0.7		-0.1	-0.5		1.0	1.7		0.3	0.3		0.3	0.2
RMM		-0.1	-0.4		1.7	1.5		1.5	1.1		1.4	1.2		1.1	0.7		1.3	1.0		2.4	2.0
COMP		4.7	3.7		25.9	28.8		26.4	30.1		25.8	30.8		21.7	26.0		31.7	36.9		19.6	20.2

Note: Solution value for REF-CC – all other scenarios show percent change from REF-CC

Table C – 33 Maize production by year and region (million metric tons)

Scenario	World			Developing			East Asia and Pacific			South Asia			Africa South of the Sahara			Middle East and North Africa			Latin America and Caribbean		
	2010	2030	2050	2010	2030	2050	2010	2030	2050	2010	2030	2050	2010	2030	2050	2010	2030	2050	2010	2030	2050
REF-CC	752	914	1023	385	538	659	177	243	301	22	31	36	45	65	72	14	17	19	98	151	197
HIGH+RE		22.0	24.6		35.5	39.3		16.7	16.0		78.2	89.9		56.4	66.2		65.0	79.2		50.3	57.4
IX+WUE		1.2	0.8		2.6	1.7		6.3	3.5		3.4	3.1		2.0	3.0		16.4	20.4		-4.3	-3.3
SWHC		1.0	2.2		1.8	3.9		1.5	3.7		1.7	2.3		4.1	8.2		0.1	0.2		1.9	4.1
RMM		3.2	3.2		5.4	5.2		4.3	4.3		7.2	6.9		5.0	4.7		3.5	2.9		7.0	6.6
COMP		30.8	35.5		49.9	56.4		31.9	30.8		102.8	117.3		76.1	97.3		98.6	120.2		58.2	70.3

Note: Solution value for REF-CC – all other scenarios show percent change from REF-CC

Table C – 34 Rice production by year and region (million metric tons)

Scenario	World			Developing			East Asia and Pacific			South Asia			Africa South of the Sahara			Middle East and North Africa			Latin America and Caribbean		
	2010	2030	2050	2010	2030	2050	2010	2030	2050	2010	2030	2050	2010	2030	2050	2010	2030	2050	2010	2030	2050
REF-CC	435.0	495.2	511.1	415.5	473.6	485.9	244.6	264.0	269.4	132.3	155.5	149.6	11.1	21.5	32.2	7.1	8.2	10.2	19.1	22.4	21.4
HIGH+RE		5.2	6.1		6.0	7.1		5.1	5.1		4.6	4.9		37.8	45.0		-0.2	-0.5		-0.6	-2.5
IX+WUE		2.0	1.8		2.3	2.1		1.7	1.0		2.4	2.6		4.2	6.2		5.6	5.3		6.2	6.3
SWHC		0.5	0.6		0.5	0.8		0.7	1.0		0.0	-0.2		2.2	3.6		-0.6	-1.3		0.1	-0.2
RMM		1.3	1.2		1.4	1.4		1.4	1.3		1.4	1.5		1.6	1.4		0.5	0.8		2.4	2.3
COMP		9.1	10.1		10.3	11.7		8.5	8.0		8.8	9.0		49.5	63.3		3.2	1.9		8.0	7.1

Note: Solution value for REF-CC – all other scenarios show percent change from REF-CC

Table C – 35 Wheat production by year and region (million metric tons)

Scenario	World			Developing			East Asia and Pacific			South Asia			Africa South of the Sahara			Middle East and North Africa			Latin America and Caribbean		
	2010	2030	2050	2010	2030	2050	2010	2030	2050	2010	2030	2050	2010	2030	2050	2010	2030	2050	2010	2030	2050
REF-CC	646.7	787.9	924.0	408.2	525.1	623.7	106.0	111.1	104.2	104.8	153.5	207.4	6.8	9.6	12.9	66.5	88.5	102.1	27.4	35.6	43.7
HIGH+RE		10.4	10.5		22.0	23.8		36.8	43.7		24.3	26.2		34.8	40.8		31.2	36.5		37.9	44.2
IX+WUE		1.4	1.4		2.9	3.0		-1.1	-1.5		12.3	11.6		4.5	4.4		0.6	0.7		-8.4	-9.0
SWHC		0.2	-0.7		0.4	-0.3		-0.0	-1.6		0.8	0.5		1.7	2.8		0.2	-0.3		0.9	0.7
RMM		1.5	0.7		3.0	2.3		2.4	1.3		3.7	2.8		2.9	2.1		2.0	1.3		4.2	3.3
COMP		14.5	15.9		29.7	33.1		38.2	44.9		45.4	48.3		47.3	58.5		35.3	41.6		33.0	40.3

Note: Solution value for REF-CC – all other scenarios show percent change from REF-CC

Table C – 36 Banana production by year and region (million metric tons)

Scenario	World			Developing			East Asia and Pacific			South Asia			Africa South of the Sahara			Middle East and North Africa			Latin America and Caribbean		
	2010	2030	2050	2010	2030	2050	2010	2030	2050	2010	2030	2050	2010	2030	2050	2010	2030	2050	2010	2030	2050
REF-CC	89.5	137.3	180.7	88.6	136.2	179.5	26.3	36.0	45.3	21.2	38.3	54.7	11.3	18.7	25.9	1.8	3.4	4.9	28.0	39.9	48.8
HIGH+RE		12.1	14.5		12.4	14.8		17.4	21.6		15.5	18.1		12.8	13.7		-16.6	-18.7		7.0	8.7
IX+WUE		0.9	1.0		0.9	1.1		-1.3	-1.5		0.4	0.3		2.6	3.6		-0.3	-1.1		2.8	3.1
SWHC		0.6	1.0		0.6	1.0		0.2	0.4		0.1	-0.3		3.4	6.1		-0.6	-1.1		0.3	0.5
RMM		1.7	1.7		1.8	1.7		2.1	2.0		2.5	2.5		1.1	1.0		0.8	0.9		1.1	1.1
COMP		15.5	18.4		15.8	18.7		18.9	22.8		19.4	21.1		20.7	26.4		-16.3	-19.9		10.0	11.9

Note: Solution value for REF-CC – all other scenarios show percent change from REF-CC

Table C – 37 Plantain production by year and region (million metric tons)

Scenario	World			Developing			East Asia and Pacific			South Asia			Africa South of the Sahara			Middle East and North Africa			Latin America and Caribbean		
	2010	2030	2050	2010	2030	2050	2010	2030	2050	2010	2030	2050	2010	2030	2050	2010	2030	2050	2010	2030	2050
REF-CC	38.9	71.0	110.9	38.9	71.0	110.9	0.8	1.4	2.2	0.7	1.2	1.8	27.6	52.3	83.6	0.0	0.0	0.0	9.8	16.1	23.3
HIGH+RE		9.8	12.0		9.8	12.0		-1.8	-2.3		11.6	14.0		11.5	14.0					4.9	5.9
IX+WUE		0.1	0.3		0.1	0.3		-0.4	-0.7		1.7	1.2		0.2	0.1					-0.1	0.8
SWHC		2.0	3.5		2.0	3.5		-0.9	-2.5		-1.2	-3.4		3.3	6.2					-1.7	-5.1
RMM		1.5	1.5		1.5	1.5		1.6	1.6		2.3	2.3		1.4	1.4					1.7	1.8
COMP		13.7	17.5		13.7	17.5		-1.7	-4.3		14.4	13.4		16.9	22.4					4.3	2.0

Note: Solution value for REF-CC – all other scenarios show percent change from REF-CC

Table C – 38 Bean production by year and region (million metric tons)

Scenario	World			Developing			East Asia and Pacific			South Asia			Africa South of the Sahara			Middle East and North Africa			Latin America and Caribbean		
	2010	2030	2050	2010	2030	2050	2010	2030	2050	2010	2030	2050	2010	2030	2050	2010	2030	2050	2010	2030	2050
REF-CC	21.2	29.3	36.8	19.1	25.9	32.2	5.6	7.0	8.3	3.1	3.4	3.1	3.4	5.1	6.7	0.6	0.7	0.7	6.1	9.3	12.9
HIGH+RE		8.1	10.1		12.8	16.1		11.1	11.9		3.4	3.4		6.3	7.7		2.7	2.9		22.2	27.6
IX+WUE		0.0	0.1		-0.2	-0.2		-0.2	-0.1		1.4	1.8		3.0	3.6		4.3	3.6		-2.9	-2.9
SWHC		0.4	0.7		0.6	1.1		0.3	0.2		-0.2	-0.5		2.2	4.5		-0.6	-1.4		0.3	0.6
RMM		1.4	1.4		2.3	2.4		2.4	2.4		3.0	3.0		2.8	2.5		4.7	4.9		2.1	2.7
COMP		10.1	12.4		15.5	19.5		14.0	14.7		7.7	7.4		15.1	19.5		11.4	8.9		21.1	27.2

Note: Solution value for REF-CC – all other scenarios show percent change from REF-CC

Table C – 39 Chickpea production by year and region (million metric tons)

Scenario	World			Developing			East Asia and Pacific			South Asia			Africa South of the Sahara			Middle East and North Africa			Latin America and Caribbean		
	2010	2030	2050	2010	2030	2050	2010	2030	2050	2010	2030	2050	2010	2030	2050	2010	2030	2050	2010	2030	2050
REF-CC	9.3	13.0	16.3	8.8	12.2	14.9	0.3	0.3	0.3	6.7	9.3	11.2	0.3	0.6	0.9	1.3	1.7	2.0	0.2	0.3	0.4
HIGH+RE		10.5	13.1		12.8	16.6		19.7	24.4		10.8	14.3		20.3	26.7		27.2	35.8		-29.3	-32.2
IX+WUE		0.5	0.6		0.6	0.7		-2.1	-2.2		-0.8	-0.9		-0.3	0.2		8.8	10.5		0.6	0.9
SWHC		0.3	0.6		0.3	0.8		0.1	0.1		0.2	0.5		1.6	3.5		0.5	1.0		0.3	0.5
RMM		1.6	1.7		2.0	2.2		1.9	2.0		2.5	2.8		2.7	2.9		0.9	1.1		-9.2	-8.8
COMP		13.4	16.4		16.1	20.8		20.2	25.7		12.9	17.0		25.2	35.4		38.6	49.2		-35.9	-38.2

Note: Solution value for REF-CC – all other scenarios show percent change from REF-CC

Table C – 40 Cowpea production by year and region (million metric tons)

Scenario	World			Developing			East Asia and Pacific			South Asia			Africa South of the Sahara			Middle East and North Africa			Latin America and Caribbean		
	2010	2030	2050	2010	2030	2050	2010	2030	2050	2010	2030	2050	2010	2030	2050	2010	2030	2050	2010	2030	2050
REF-CC	5.5	9.5	15.4	5.4	9.4	15.3	0.2	0.2	0.2	0.0	0.0	0.0	5.1	9.1	14.9	0.0	0.0	0.0	0.1	0.1	0.2
HIGH+RE		15.2	19.3		15.5	19.7		-15.3	-18.1		-17.8	-20.8		16.8	21.1		-24.7	-27.9		-27.1	-30.8
IX+WUE		0.3	0.3		0.3	0.3		-1.3	-1.3		-0.9	-1.0		0.3	0.3		3.2	3.0		2.9	3.2
SWHC		1.2	2.1		1.2	2.1		-0.4	-1.4		0.9	1.2		1.3	2.3		-0.7	-2.2		-1.2	-3.6
RMM		2.6	2.6		2.7	2.7		0.5	0.5		3.0	2.9		2.8	2.8		1.5	1.6		0.1	0.2
COMP		19.7	24.9		20.1	25.4		-16.6	-20.3		-15.3	-18.6		21.7	27.0		-21.8	-26.6		-26.8	-32.2

Note: Solution value for REF-CC – all other scenarios show percent change from REF-CC

Table C – 41 Lentil production by year and region (million metric tons)

Scenario	World			Developing			East Asia and Pacific			South Asia			Africa South of the Sahara			Middle East and North Africa			Latin America and Caribbean		
	2010	2030	2050	2010	2030	2050	2010	2030	2050	2010	2030	2050	2010	2030	2050	2010	2030	2050	2010	2030	2050
REF-CC	3.9	5.5	6.8	2.6	3.1	3.2	0.1	0.2	0.2	1.3	1.5	1.6	0.1	0.1	0.2	1.0	1.2	1.2	0.02	0.03	0.03
HIGH+RE		8.5	9.8		30.0	41.2		19.2	26.5		29.9	41.4		16.1	23.8		33.7	47.0		8.1	15.3
IX+WUE		0.5	0.5		1.1	1.2		-0.6	-0.6		-0.4	-0.2		-0.7	-0.7		3.6	4.0		-4.0	-3.6
SWHC		0.2	0.5		0.5	1.1		0.5	1.1		0.4	1.0		1.8	4.5		0.3	0.6		0.9	1.9
RMM		1.2	1.2		4.2	4.8		4.3	4.6		3.7	4.3		5.7	6.2		5.0	5.6		0.0	1.1
COMP		10.8	12.2		36.7	50.1		24.2	33.6		34.1	47.7		23.4	35.9		44.5	60.2		2.1	11.9

Note: Solution value for REF-CC – all other scenarios show percent change from REF-CC

Table C – 42 Pigeonpea production by year and region (million metric tons)

Scenario	World			Developing			East Asia and Pacific			South Asia			Africa South of the Sahara			Middle East and North Africa			Latin America and Caribbean		
	2010	2030	2050	2010	2030	2050	2010	2030	2050	2010	2030	2050	2010	2030	2050	2010	2030	2050	2010	2030	2050
REF-CC	3.8	5.5	6.9	3.8	5.5	6.9	0.6	0.8	0.9	2.7	4.1	5.2	0.4	0.6	0.7	0.0	0.0	0.0	0.04	0.07	0.10
HIGH+RE		11.6	14.7		11.6	14.7		-14.3	-17.0		19.0	23.2		0.0	1.2					-30.4	-35.1
IX+WUE		0.1	0.2		0.1	0.2		-1.1	-1.1		0.3	0.3		0.9	1.2					0.3	0.6
SWHC		0.3	0.7		0.3	0.7		0.4	0.7		0.0	0.2		2.0	3.8					0.7	1.1
RMM		1.7	1.8		1.7	1.8		1.3	1.3		1.8	1.9		1.9	1.9					1.5	1.5
COMP		14.2	18.0		14.2	18.0		-13.7	-16.1		21.7	26.2		5.1	8.8					-28.5	-32.7

Note: Solution value for REF-CC – all other scenarios show percent change from REF-CC

Table C – 43 Groundnut production by year and region (million metric tons)

Scenario	World			Developing			East Asia and Pacific			South Asia			Africa South of the Sahara			Middle East and North Africa			Latin America and Caribbean		
	2010	2030	2050	2010	2030	2050	2010	2030	2050	2010	2030	2050	2010	2030	2050	2010	2030	2050	2010	2030	2050
REF-CC	36.5	46.1	50.4	34.7	42.9	47.0	16.5	19.6	20.6	6.9	7.2	6.3	9.6	14.0	17.8	0.4	0.5	0.5	1.2	1.5	1.7
HIGH+RE		6.6	8.7		8.9	11.3		8.3	9.8		9.5	11.2		14.2	17.7		-21.1	-23.8		-22.9	-25.4
IX+WUE		0.5	0.6		0.5	0.7		-0.4	-0.8		1.2	1.0		2.2	3.1		2.3	2.1		-7.0	-7.7
SWHC		0.4	0.7		0.5	0.9		0.1	-0.1		0.4	0.6		1.1	2.1		-0.4	-1.0		0.5	1.0
RMM		-0.5	-0.4		-0.8	-0.8		-0.9	-1.0		-1.1	-1.3		-0.4	-0.3		-2.3	-2.2		-0.7	-0.5
COMP		7.0	9.4		9.2	12.0		6.8	7.3		9.6	10.5		17.6	23.4		-21.3	-24.9		-28.4	-31.2

Note: Solution value for REF-CC – all other scenarios show percent change from REF-CC

Table C – 44 Soybean production by year and region (million metric tons)

Scenario	World			Developing			East Asia and Pacific			South Asia			Africa South of the Sahara			Middle East and North Africa			Latin America and Caribbean		
	2010	2030	2050	2010	2030	2050	2010	2030	2050	2010	2030	2050	2010	2030	2050	2010	2030	2050	2010	2030	2050
REF-CC	231.2	304.6	338.9	135.1	186.6	210.9	19.8	25.9	28.0	8.7	10.8	11.2	1.3	1.6	1.5	0.3	0.3	0.3	103.1	145.6	167.5
HIGH+RE		0.1	0.4		-1.3	-1.7		-0.7	-0.9		0.0	-0.1		32.8	37.2		-3.1	-0.8		-1.9	-2.4
IX+WUE		-0.2	-0.1		-1.5	-1.4		2.2	2.2		2.6	3.3		2.3	2.9		19.3	31.6		-2.6	-2.5
SWHC		0.4	0.6		0.8	1.4		0.6	0.9		0.4	0.3		3.8	6.4		2.0	0.3		0.9	1.5
RMM		0.0	0.1		-1.0	-1.0		-0.9	-1.2		-1.7	-2.0		-1.2	-1.2		-2.4	-1.4		-0.9	-0.9
COMP		0.2	0.7		-2.9	-2.8		1.1	0.9		1.1	1.3		39.3	48.4		16.8	28.1		-4.6	-4.3

Note: Solution value for REF-CC – all other scenarios show percent change from REF-CC

Table C – 45 Palm fruit production by year and region (million metric tons)

Scenario	World			Developing			East Asia and Pacific			South Asia			Africa South of the Sahara			Middle East and North Africa			Latin America and Caribbean		
	2010	2030	2050	2010	2030	2050	2010	2030	2050	2010	2030	2050	2010	2030	2050	2010	2030	2050	2010	2030	2050
REF-CC	229.7	439.7	609.6	229.7	439.7	609.6	183.8	361.0	517.4	0.0	0.0	0.0	35.4	61.6	73.9	0.0	0.0	0.0	10.4	17.1	18.2
HIGH+RE		2.9	4.0		2.9	4.0		3.4	4.3					1.1	2.5					0.0	2.1
IX+WUE		0.3	0.3		0.3	0.3		0.2	0.1					0.9	1.4					0.8	1.6
SWHC		0.3	0.5		0.3	0.5		-0.1	-0.3					3.0	6.0					-0.1	0.3
RMM		0.1	0.0		0.1	0.0		0.2	0.2					-0.3	-0.6					-0.8	-1.3
COMP		3.6	4.6		3.6	4.6		3.6	4.2					4.6	8.6					-0.2	2.1

Note: Solution value for REF-CC – all other scenarios show percent change from REF-CC

Table C – 46 Food oil production by year and region (million metric tons)

Scenario	World			Developing			East Asia and Pacific			South Asia			Africa South of the Sahara			Middle East and North Africa			Latin America and Caribbean		
	2010	2030	2050	2010	2030	2050	2010	2030	2050	2010	2030	2050	2010	2030	2050	2010	2030	2050	2010	2030	2050
REF-CC	134.2	198.0	242.6	104.1	160.6	202.4	58.2	97.5	130.8	9.0	11.3	11.6	9.0	14.4	17.1	3.0	3.8	4.3	19.0	26.5	30.1
HIGH+RE		1.5	2.3		1.5	2.2		2.5	3.3		1.3	1.7		0.7	1.9		-0.8	-1.3		-1.3	-1.5
IX+WUE		0.2	0.2		0.1	0.2		0.1	0.0		2.2	2.8		0.9	1.4		1.6	1.8		-1.3	-1.3
SWHC		0.2	0.4		0.3	0.4		-0.0	-0.1		0.4	0.8		2.1	4.3		0.2	0.6		0.4	0.6
RMM		0.0	0.0		-0.3	-0.3		-0.1	-0.1		-0.9	-1.3		-0.3	-0.5		-0.9	-0.9		-0.4	-0.5
COMP		1.8	2.6		1.6	2.4		2.3	3.0		3.1	3.7		3.4	6.7		0.1	-0.2		-2.6	-2.8

Note: Solution value for REF-CC – all other scenarios show percent change from REF-CC

Table C – 47 Fruit and vegetable production by year and region (million metric tons)

Scenario	World			Developing			East Asia and Pacific			South Asia			Africa South of the Sahara			Middle East and North Africa			Latin America and Caribbean		
	2010	2030	2050	2010	2030	2050	2010	2030	2050	2010	2030	2050	2010	2030	2050	2010	2030	2050	2010	2030	2050
REF-CC	1592	2290	2913	1305	1919	2447	670	901	1077	159	300	411	101	173	258	140	230	329	164	224	271
HIGH+RE		0.8	1.9		0.0	1.0		-0.6	0.4		0.8	1.5		3.5	5.0		-2.5	-2.1		-0.7	0.3
IX+WUE		0.5	0.7		0.6	0.9		-1.0	-1.1		-0.7	-0.9		3.0	4.7		3.1	2.5		4.3	5.4
SWHC		0.3	0.7		0.4	0.8		0.3	0.6		-0.1	-0.2		2.4	4.7		0.0	0.2		0.3	0.6
RMM		1.6	1.7		2.5	2.6		3.7	3.7		4.6	4.6		1.0	1.0		1.7	2.2		0.9	1.1
COMP		3.1	4.8		3.5	5.2		2.6	3.8		4.7	5.1		9.8	15.6		2.3	2.3		4.3	6.7

Note: Solution value for REF-CC – all other scenarios show percent change from REF-CC

Income, kilocalories, and food security

Table C – 48 Income by year and region (per capita GDP in thousands 2005 US$)

Scenario	World			Developing			East Asia and Pacific			South Asia			Africa South of the Sahara			Middle East and North Africa			Latin America and Caribbean		
	2010	2030	2050	2010	2030	2050	2010	2030	2050	2010	2030	2050	2010	2030	2050	2010	2030	2050	2010	2030	2050
REF-CC	9.8	17.2	24.7	5.4	12.4	19.5	8.8	22.3	35.1	2.7	6.8	13.0	2.0	3.7	7.2	10.0	17.0	25.7	10.0	16.9	25.6
HIGH+RE		1.1	2.4		1.7	3.4		1.1	1.9		2.7	4.5		6.5	13.4		1.2	2.1		0.5	1.0
IX+WUE		0.1	0.2		0.2	0.3		0.2	0.1		0.3	0.3		0.3	0.9		0.2	0.3		0.2	0.3
SWHC		0.1	0.3		0.1	0.5		0.1	0.3		0.1	0.2		0.7	2.4		0.0	0.2		0.0	0.1
RMM		0.2	0.4		0.4	0.5		0.3	0.4		0.9	1.2		0.4	0.7		0.1	0.2		0.1	0.1
COMP		1.5	3.2		2.4	4.6		1.6	2.6		3.9	6.2		7.7	16.9		1.6	2.8		0.8	1.5

Note: Solution value for REF-CC – all other scenarios show percent change from REF-CC

Table C – 49 Average calorie availability by year and region (kilocalories per person per day)

Scenario	World			Developing			East Asia and Pacific			South Asia			Africa South of the Sahara			Middle East and North Africa			Latin America and Caribbean		
	2010	2030	2050	2010	2030	2050	2010	2030	2050	2010	2030	2050	2010	2030	2050	2010	2030	2050	2010	2030	2050
REF-CC	2794	2978	3066	2682	2904	3004	2876	3245	3310	2360	2613	2815	2356	2507	2683	3124	3207	3278	2876	2984	3083
HIGH+RE		3.8	5.1		4.2	5.6		3.4	4.3		4.8	6.1		6.7	9.0		3.3	4.2		2.3	2.9
IX+WUE		0.7	0.6		0.7	0.7		0.8	0.7		0.9	0.9		0.6	0.7		0.6	0.6		0.4	0.4
SWHC		0.3	0.7		0.3	0.7		0.3	0.6		0.3	0.6		0.6	1.3		0.2	0.5		0.2	0.4
RMM		0.0	0.1		0.1	0.1		0.0	0.0		0.1	0.2		0.1	0.1		0.0	0.0		0.0	0.0
COMP		4.5	5.9		5.0	6.5		4.2	5.1		5.9	7.2		7.5	10.0		3.9	4.8		2.7	3.3

Note: Solution value for REF-CC – all other scenarios show percent change from REF-CC

Table C – 50 Population at risk of hunger by year and region (millions)

Scenario	World			Developing			East Asia and Pacific			South Asia			Africa South of the Sahara			Middle East and North Africa			Latin America and Caribbean		
	2010	2030	2050	2010	2030	2050	2010	2030	2050	2010	2030	2050	2010	2030	2050	2010	2030	2050	2010	2030	2050
REF-CC	838.1	602.6	491.0	823.3	587.0	475.4	271.3	119.9	108.8	268.5	167.3	99.1	209.5	226.8	199.7	29.3	35.8	38.3	39.5	35.9	28.8
HIGH+RE		-24.2	-21.5		-24.6	-21.6		-11.4	-14.4		-37.1	-14.8		-26.6	-31.6		-11.6	-13.6		-12.8	-15.0
IX+WUE		-4.2	-3.2		-4.3	-3.3		-3.3	-2.5		-8.1	-3.6		-2.6	-3.9		-2.3	-2.2		-2.9	-2.7
SWHC		-2.1	-3.7		-2.2	-3.8		-1.2	-1.8		-2.6	-2.4		-2.8	-6.1		-0.8	-1.9		-1.3	-2.5
RMM		-0.4	-0.2		-0.4	-0.2		-0.1	-0.1		-1.2	-0.3		-0.2	-0.4		0.0	0.0		0.1	0.1
COMP		-27.7	-23.5		-28.2	-23.7		-13.9	-16.3		-44.0	-16.3		-28.7	-34.4		-13.4	-15.0		-15.0	-16.9

Note: Solution value for REF-CC – all other scenarios show percent change from REF-CC

Table C – 51 Undernourished children by year and region (millions of undernourished children, by weight, 5 years old and younger)

Scenario	World			Developing			East Asia and Pacific			South Asia			Africa South of the Sahara			Middle East and North Africa			Latin America and Caribbean		
	2010	2030	2050	2010	2030	2050	2010	2030	2050	2010	2030	2050	2010	2030	2050	2010	2030	2050	2010	2030	2050
REF-CC	150.1	128.3	105.8	149.8	128.1	105.6	21.9	11.9	8.3	77.1	64.8	52.3	41.0	44.5	39.9	4.0	2.8	1.9	4.3	3.0	1.8
HIGH+RE		-5.1	-7.1		-5.1	-7.1		-8.0	-6.0		-3.1	-4.3		-6.5	-10.1		-12.5	-15.1		-8.6	-14.8
IX+WUE		-0.8	-0.8		-0.8	-0.8		-1.9	-1.1		-0.6	-0.6		-0.6	-0.8		-2.3	-2.2		-1.7	-2.3
SWHC		-0.4	-1.0		-0.4	-1.0		-0.7	-0.8		-0.2	-0.5		-0.6	-1.5		-0.7	-1.8		-0.8	-2.2
RMM		-0.1	-0.1		-0.1	-0.1		-0.1	-0.1		-0.1	-0.1		-0.1	-0.1		0.0	0.0		0.0	0.1
COMP		-5.9	-8.0		-5.9	-8.0		-10.0	-7.2		-3.7	-5.1		-7.1	-11.2		-14.9	-17.0		-10.1	-16.5

Note: Solution value for REF-CC – all other scenarios show percent change from REF-CC

Table C – 52 Comparing results to World Health Organization (WHO) targets for a healthy diet in 2010, 2030, and 2050, by region and scenario

Region	2010				Scenario	2030				2050			
	F&V[1]	Fat[2]	Sugar[3]	KCAL[4]		F&V	Fat	Sugar	KCAL	F&V	Fat	Sugar	KCAL
East Asia and Pacific	813	22.1	9.2	2,876	**REF-CC**	**979**	**25.8**	**10.5**	3,245	**963**	**27.5**	**11.1**	3,310
					HIGH+RE	978	25.5	10.2	3,355	963	27.1	10.7	3,452
					IX+WUE	983	25.6	10.4	3,271	968	27.4	11.0	3,333
					SWHC	981	25.8	10.5	3,255	967	27.5	11.1	3,330
					RMM	993	25.7	10.5	3,245	977	27.4	11.0	3,310
					COMP	998	25.2	10.1	3,381	983	26.8	10.6	3,479
South Asia	287	17.0	11.4	2,360	**REF-CC**	**526**	**18.3**	**14.7**	2,613	**939**	**18.8**	**16.9**	2,815
					HIGH+RE	539	18.0	14.3	2,738	979	18.4	16.3	2,987
					IX+WUE	532	18.2	14.7	2,637	952	18.7	16.9	2,840
					SWHC	529	18.3	14.8	2,621	949	18.7	16.9	2,832
					RMM	539	18.2	14.7	2,616	961	18.6	16.9	2,821
					COMP	560	17.9	14.2	2,767	1020	18.3	16.2	3,018
Africa South of the Sahara	262	16.2	8.4	2,356	**REF-CC**	**320**	**16.9**	**8.9**	2,507	**391**	**18.1**	**9.8**	2,683
					HIGH+RE	332	16.8	8.7	2,675	410	18.0	9.4	2,924
					IX+WUE	321	16.9	8.9	2,522	393	18.1	9.8	2,702
					SWHC	322	16.9	8.9	2,522	396	18.1	9.8	2,718
					RMM	325	16.8	8.9	2,510	397	18.1	9.7	2,686
					COMP	341	16.7	8.7	2,695	423	17.8	9.4	2,951
Middle East and North Africa	752	20.0	13.5	3,124	**REF-CC**	**778**	**20.6**	**15.1**	3,207	**784**	**21.4**	**17.0**	3,278
					HIGH+RE	772	20.4	14.7	3,313	777	21.1	16.4	3,416
					IX+WUE	780	20.5	15.1	3,226	787	21.3	17.0	3,298
					SWHC	779	20.6	15.1	3,213	787	21.3	17.0	3,294
					RMM	787	20.5	15.1	3,207	793	21.3	16.9	3,278
					COMP	784	20.3	14.7	3,332	790	21.0	16.4	3,435
Latin America and Caribbean	469	24.7	18.2	2,876	**REF-CC**	**492**	**25.5**	**19.0**	2,984	**536**	**26.6**	**19.8**	3,083
					HIGH+RE	497	25.4	18.6	3,053	543	26.4	19.2	3,172
					IX+WUE	494	25.5	19.0	2,996	538	26.6	19.8	3,095
					SWHC	494	25.5	19.0	2,990	539	26.6	19.7	3,095
					RMM	499	25.5	19.0	2,984	544	26.6	19.7	3,083
					COMP	508	25.3	18.6	3,065	557	26.3	19.2	3,185

Heatmap Legend

Failed to achieve target **Achieved target** **Surpassed target**

Notes:
[1] In g/person/day; WHO healthy diet recommendation: at least 400 g/person/day of fruits and vegetables
[2] In percent (%); WHO healthy diet recommendation: less than 30% of energy from fat
[3] In percent (%); WHO healthy diet recommendation: less than 10% of energy from sugar, less than 5% ideal
[4] In kcal/person/day, WHO healthy diet recommendation: energy intake should correspond with activity levels; 3,000 kcal/person/day and 2,400 kcal/person/day for active 20–35-year-old males and females, respectively

Appendix D. Further methods on hunger calculations and scenario specifications

Prevalence of undernourishment and hunger calculation

(adapted from Robinson et al. 2015)

The Prevalence of Undernourishment is a population measure to estimate the share at risk of suffering from chronic hunger. This calculation is based on a strong empirical correlation between the share of undernourished within the total population and the relative availability of food and is adapted from the *FAO Methodology for the Measurement of Food Deprivation Updating the Minimum Dietary Energy Requirements* (FAO 2008).

$$ShareAtRisk = \alpha RelativeKCal^2 + \beta RelativeKCal + int + \varepsilon$$

$$\alpha = 89.63$$

$$\beta = -319.69$$

$$\text{int} = 288.16$$

$$\varepsilon = \text{Estimation error}$$

$$RelativeKCal = \frac{KCal}{MinKCal}$$

$$Kcal = \text{Food supply}$$

$$MinKcal = \text{Minimum food requirement}$$

It should be noted that due to the quadratic nature of this equation it is necessary to apply an upper and lower bound to the share at risk. The lower bound is defined as 0 and the upper bound is 100. Developed countries unsurprisingly have a low share at risk, so for simplicity we treat all countries with less than a 4 percent share at risk of hunger as if they had a 0 percent share of hunger. The relative availability of food was bounded to ensure realistic results on the quadratic curve: when the ratio of calories available to calories required, RelativeKCal, is greater than 1.7, we assume that the share at risk of hunger is effectively 0.

Scenario specifications

(summarized from full description in Rosegrant et al. 2017)

Reference Scenario: Agricultural research

Improvements in agricultural productivity in the reference scenario are represented by exogenous growth rates for each commodity and country, based on historical trends as well as expert opinion about future changes. We developed an R&D investment-yield model to assess the investment required to achieve projected growth in agricultural productivity.

Investments in research take time to bear fruit, as innovative ideas can take years to be developed and diffused widely. To capture these lags, the investment-yield estimation model is based on the perpetual inventory method (PIM), where research investments contribute to the stock of knowledge over time. Knowledge decays as older technologies become obsolete or irrelevant. Productivity grows if the stock of knowledge grows at a faster rate than the stock of knowledge decays. The lag structure in the PIM used here follows a gamma distribution, where R&D investments reach peak impact 10 years after initial investment and then decline over time to reach zero 10 years after peak impact. With regionally differentiated research elasticities and decay rates, these imputed lag structures would vary by region according to existing R&D capacity and the potential trajectories for each region.

Research capacity varies significantly by region. To reflect these differences, we use elasticities of productivity with respect to research investments from the literature and incorporate spillover effects to represent regional capacity to access and apply outside knowledge. To meet the productivity growth trajectories assumed in the reference scenario, required investment in agricultural R&D by the CGIAR is projected to average US$1.7 billion per year between 2015 and 2050 in real 2005 dollars, while annual national agricultural research systems (NARS) investment in developing countries averages US$6.4 billion (Table D – 1). The largest investments are projected in Africa south of the Sahara (SSA) (US$2.2 billion per year) and Latin America and the Caribbean (LAC) (US$1.8 billion per year). In most regions, the larger contribution to agricultural research will come from investments from NARS, except for SSA, where about one-half of the investments will come from the CGIAR.

Table D – 1. Average annual investments in developing countries in the reference scenario (REF_CC), 2015–2050 (billion 2005 US$)

Region	R&D			Water			Infrastructure			
	CGIAR	NARS	Total	Irrig.	WUE	SWHC	Road	Rail	Elect.	Total
EAP	0.07	1.54	1.60	1.29	0.94	0.34	5.76	1.63	6.63	14.02
SAS	0.26	0.71	0.97	1.82	0.76	0.17	2.34	2.39	2.81	7.54
SSA	1.11	1.11	2.23	2.98	0.13	0.39	0.02	0.02	0.14	0.18
MEN	0.09	1.41	1.50	0.81	0.07	0.11	0.05	0.02	0.84	0.91
LAC	0.20	1.59	1.78	0.66	0.31	0.28	0.06	0.13	0.58	0.77
DVG	1.73	6.36	8.08	7.56	2.21	1.29	8.23	4.19	11.00	23.42

Notes: Figures are average annual investments over 2015–2050. R&D-Research and Development; Irrig-Irrigation; WUE- Water Use Efficiency; SWHC-Investments in Soil-Water Management; Elect-Electricity. Regions are EAP-East Asia and Pacific; SAS-South Asia; SSA-Africa south of the Sahara; MEN-Middle East and North Africa; LAC-Latin America and the Caribbean; DVG-Developing World. Regions only include developing countries.

Reference Scenario: Irrigation and water resource management

Water availability, including rainfall, stream flows, and evaporation, is determined in a hydrological model that downscales precipitation and temperature from climate scenarios generated by the Global Circulation Model (GCM). Water supply and demand by sector are determined in a water simulation model that allocates water across irrigation, livestock, domestic use, and industrial use. Water supply and demand are solved in 154 river basins globally and are linked annually to the IMPACT economic model (Robinson et al. 2015). Two of the key drivers in this model are assumptions on trends in irrigation expansion and water use efficiency (WUE). Similar to assumptions on agricultural productivity, the reference assumptions used for these drivers are based on historical trends combined with expert opinion about future pathways. Total harvested area expands by about 18 percent in the projection period from 2010 to 2050. Irrigated area grows at a faster rate than rainfed area. Expansion of irrigation requires investments in water infrastructure such as dams, canals, and other conveyance systems. Across all developing countries, the cost of investments required to achieve the projected irrigation expansion is US$7.6 billion per year, with the cost per hectare of new irrigation systems taken from literature. While the largest expansion in irrigated area is projected in Asia, the largest investments will be needed in Africa south of the Sahara (SSA) due to the higher costs of expanding irrigation.

Similar to the cost of expanding irrigation, the cost for increased WUE is estimated at the hectare level and specified by region. Investments in improved WUE across the developing world are less than one-third of the investments projected for irrigation expansion, at US$2.2 billion per year. The majority of these investments are projected in East Asia and Pacific (EAP) and South Asia (SAS), which account for almost 80 percent and 77 percent of irrigated area in developing countries in 2010 and 2050, respectively. Baseline investments in soil water management (SWHC) technologies are from limited evidence but based on previous studies we arrive at a set of estimates for the developing world at US$1.3 billion per year.

Reference Scenario: Infrastructure

The economic growth assumed in the reference scenario also includes investments in new and maintenance of existing infrastructure, such as electrification and roads. Based on the methodology in Rosegrant et al. (2015), we project reference investments across developing countries totaling US$23.4 billion per year.

Investment Scenarios: R&D for enhanced productivity

The HIGH+RE scenario incorporates yield gains from increasing investments in CGIAR R&D and was developed in collaboration with all 15 CGIAR Centers through the Global Futures and Strategic Foresight program. As a starting point, each Center quantified potential yield gains for its respective commodities (including crops, livestock, and fish) in developing countries across major regions of the world (SAS, SSA, LAC, EAP, and MENA) with increased R&D investment. These yield gains were first expressed as potential changes in absolute yield levels and then translated into differential yield growth rates used in the IMPACT modeling framework. The final endogenous yields and output growth generated by the scenarios are functions of interactions between these growth rates and projected changes in prices, demand, and other factors. Increased CGIAR research efficiency accelerates and augments productivity gains so that the yield impact of investments is 30 percent higher and the maximum improvement is achieved earlier than in a business-as-usual (BAU) situation. Research efficiency is gained through advancement in breeding techniques including further advances in genomics and bio-informatics and high throughput gene sequencing; and more effective regulatory and intellectual property rights systems that reduce the lag times from discovery to deployment of new varieties. Table D – 2 summarizes the additional investment needs for the productivity enhancement and other alternative investment scenarios, incremental to the reference costs for REF_CC in Table D – 1. The additional investment costs associated with these scenarios vary by region and scenario.

Investment Scenarios: Improved water resource management

Two alternative scenarios focus on investments and improvements in agricultural water resource management that affect crops and livestock directly through changes in the availability of water, and also livestock indirectly through changes in feed prices. They include accelerated investments in irrigation expansion combined with improved water use efficiency on all irrigated cropland (IX+WUE); and a focus on improvements in rainwater harvesting and soil water holding capacity (SWHC). These scenarios were developed in consultation with the International Water Management Institute (IWMI).

Table D – 2. Average annual additional investments in developing countries (relative to the reference scenario), 2015–2050 (billion 2005 US$)

Region	Scenarios				
	HIGH+RE	IX+WUE	SWHC	RMM	COMP
EAP	0.02	2.86	0.68	3.81	7.36
SAS	0.14	1.65	0.95	2.20	4.94
SSA	1.66	1.42	1.20	1.90	6.18
MENA	0.04	0.65	0.88	0.86	2.43
LAC	0.11	1.53	0.87	2.04	4.55
DVG	1.97	8.10	4.58	10.81	25.46

Notes: Figures are average annual investments over 2015–2050. Regions are EAP-East Asia and Pacific; SAS-South Asia; SSA-Africa south of the Sahara; MENA-Middle East and North Africa; LAC-Latin America and the Caribbean; DVG-Developing World. Regions only include developing countries.

The IX+WUE scenario simulates an expansion of irrigated areas in developing countries by 2030, relative to the REF_CC scenario, with changes thereafter following the growth rates in the REF_CC scenario. Irrigation expansion in the IX+WUE scenario is designed to have a neutral effect on total agricultural land across developing countries, with projected irrigated expansion of 20 million hectares roughly offset by a reduction of 22 million hectares of rainfed agriculture. In some regions, irrigation expansion will not completely displace rainfed area and will result in an overall expansion of land, as is the case in SSA, where total harvested area increases by 2.7 million and 4.3 million hectares compared to REF_CC by 2030 and 2050, respectively.

In IX+WUE, improving irrigation efficiency can increase agricultural output while conserving more water. Significant variation arises globally regarding the efficiency of water systems, ranging from highly efficient systems based on high efficiency technologies (for example, sprinklers and drip irrigation), to less efficient systems, such as furrow irrigation, and these system-level efficiencies do not necessarily translate to river basin efficiencies. IMPACT uses the concept of basin efficiency, defined as the ratio of beneficial water depletion (crop evapotranspiration and salt leaching) to total irrigation water depletion at the basin scale. Basin efficiency in future years is assumed to increase at a prescribed rate in a food production unit (FPU) depending on water infrastructure investment and water management improvement in the FPU. For the WUE scenario, basin efficiencies are assumed to increase by 15 percentage points by 2030 and then continue on previous trajectories.

The SWHC scenario simulates the benefits of technologies – such as no-till agriculture and water harvesting – that increase soil water holding capacity or otherwise make precipitation more readily available to plants (that is, effective precipitation). Improvements vary by region and represent the different levels at which these kinds of technologies are currently being applied in various regions, with a maximum increase in effective precipitation of 5–15 percent by 2045.

Each of these types of agricultural water management interventions are associated with different investment costs. These interventions include the cost of developing new technologies and infrastructure (canals, dams, etc.), as well as efforts to increase the capacity for water management. Across all developing countries, the additional cost of irrigation expansion amounts to US$3.5 billion per year. Because of differences in investment costs and assumed expansion of irrigated area, the cost of developing additional irrigation infrastructure is largest in SSA (US$1.2 billion per year). The region contributes about one-eighth of the projected irrigation expansion in hectares, but accounts for about one-third of the additional investment costs (Table D – 2).

For the IX+WUE scenario, the investment cost is composed of two portions, irrigation expansion plus improvements in water use efficiency. Improving WUE is about 30 percent more expensive than solely expanding irrigation using conventional methods. This is in large part because these improvements are implemented across a larger area, as the improved average water use efficiency is for all irrigated area (412 million hectares across developing countries), not just the newly expanded area (20 million hectares). The IX+WUE scenario requires US$8.1 billion per year in additional investments across developing countries.

The cost of implementing measures to enhance soil water capacity for crop production (SWHC) is estimated by applying a cost per hectare (US$179) to both rainfed and irrigated cropland across developing regions. The cost for the SWHC scenario is US$4.6 billion per year across developing countries.

Investment Scenarios: Improved infrastructure and market access

The Reduced Marketing Margins (RMM) scenario assumes a mix of infrastructure improvements throughout the economies of developing countries, focusing primarily on improvements to transportation infrastructure (road building, road maintenance, and railroads) and increased rural electrification. These improvements enhance productivity along the value chain, increase the speed of moving commodities to markets, as well as improve storage capacity, all of which improve market efficiency by better matching supply and demand over time. These improvements are represented as a reduction in the cost of moving goods from the farm to market. In IMPACT, this is done by adjusting the price wedges between producer and consumer prices, reducing the margin from producer prices to consumer prices by 1 percentage point per year between 2015 and 2030. The average annual investment from 2015 to 2050 is US$10.8 billion across all developing countries.

Investment Scenarios: Comprehensive investment portfolio

The final alternative investment scenario considers the potential outcomes of a comprehensive investment portfolio (COMP) that combines the investments of four of the scenarios previously specified, representing a broad array of investments across different parts of the agriculture system. These include one of the productivity enhancement scenarios (HIGH+RE), two of the improved water management scenarios (IX+WUE and SWHC), and the investments in infrastructure to reduce marketing inefficiencies (RMM). The total additional cost for the COMP scenario is US$25.5 billion per year from 2015 to 2050.

Made in United States
Troutdale, OR
11/03/2026

56844703R00036